THE MICROBIOLOGY COMPANION
SECOND EDITION REVISED
a study guide for students by students

Joel Topf M.D.

Sarah Faubel M.D.

Second Edition, Revised, Copyright 1997
    Second Printing: November, 1998
    First Printing: August, 1997

Second Edition, Copyright 1994
First Edition, Copyright 1993

ISBN: 0-9640124-1-3

Notice: the authors and Alert and Oriented Publishing Co. have taken care to make certain that the information provided in this book is as up to date as possible. However, the field of medicine is constantly changing, and we encourage the reader to consult the most up-to-date sources before treating patients. This book is intended to help with the class, medical microbiology, and it is **not intended to be a treatment guide.**

Send all inquiries to:

    Alert & Oriented Publishing Co.
    7850 El Paseo Grande, #5
    La Jolla, CA 92037

Call toll-free:
    1 (888) ALERT-44

Contact us at our website:
    www.ALERTandONLINE.com

Introduction IX

What's new X

Acknowledgments XI

Abbreviations used in the text..........................XII

Section one: Bacteria   1

Gram Positive Cocci

Staphylococcus
epidermidis..3
saprophyticus...3
aureus..4

Streptococcus
pyogenes, group A 6
agalactiae, group B..9
pneumoniae, pneumococcus 10
viridans group 10
bovis, group D 11

Enterococcus
faecalis..11
faecium 11

Gram Negative Cocci

Neisseria
meningitidis, meningococcus...........................12
gonorrhoeae, gonococcus 13

Gram Positive Rods

Bacillus
anthracis 17
cereus 17

Clostridium
botulinum...20
tetani...22
    tetanus immunization 23
perfringens 24
Clostridium difficile 24

Corynebacterium
diphtheriae...26

Listeria
monocytogenes...27

Gram Negative Rods

Enterics
Enterobacteriaceae
    Escherichia coli.......................................36
        Enterotoxigenic
        Enteroinvasive
        Enteropathogenic
        Enterohemorrhagic
    Shigella..38
        dysenteriae
        flexneri
        boydii
        sonnei
    Salmonella..39
        typhi
        cholerae suis
        paratyphi
    Klebsiella
        pneumonia..41
    Enterobacter
        cloacae 41
    Serratia
        marcescens 41
    Proteus...41
    Providencia 41
    Morganella 41

Table of Contents   The Microbiology Companion, Topf and Faubel ©1997   **III**

| Gram + Cocci | Gram – Cocci | Gram + Rods | Gram – Rods | Gram Nothing | DNA Viruses | RNA Viruses | Fungi | Parasites | X–Reference | Index |

Table of Contents

Pseudomonadaceae
Pseudomonas
aeruginosa 43
Vibrionaceae
Vibrio
cholera 44
parahaemolyticus 44
Campylobacteraceae
Campylobacter
jejuni 46
Helicobacter
pylori 47

Anaerobes
Bacteroides
fragilis 48
Prevotella
melaninogenica 48
Fusobacterium 48

Respiratory Infections
Haemophilus
influenza 50
aegyptius 51
ducreyi 51
Legionella
pneumophila 52
Bordetella
pertussis 52

Zoonotic Bacilli
Brucella 54
abortus
melitensis
suis
Francisella
tularensis 55
Yersinia
pestis 56
enterocolitica 56

Spirochetes
Borrelia
burgdorferi 59
recurrentis 59
Leptospira
interrogans 60
Treponema
pallidum 61
Laboratory diagnosis of syphilis 63

Obligate Intracellular Parasites
Chlamydia
pneumonia 68
psittaci 68
trachomatis 69
Rickettsiae
rickettsii 73
akari 73
prowazekii 74
typhi 75
tsutsugamushi 75
Coxiella
burnetii 76

Lack a Cell Wall
Mycoplasma
pneumoniae 78

Actinomycetes
Actinomyces
israelii 79
Nocardia
asteroides 79

Acid Fast Bacilli

Mycobacteria

tuberculosis 81
    Natural history of tuberculosis 84
leprae 86

Atypical Mycobacteria

kansasii 88
marinum 88
scrofulaceum 88
avium-intracellulare complex 88
fortuitum complex 88

Section two: Viruses 91

DNA Viruses

Enveloped DNA Viruses

Hepatitis overview 94
Hepadnavirus
    Hepatitis B Virus 95
        Hepatitis B Diseases 97
        HBV Antigens and Antibodies 99
    Hepatitis delta virus 101
Poxviridae
    Smallpox virus 104
    Vaccinia 104
    Molluscum contagiosum virus 104
Herpesviridae
    Herpes simplex virus I 108
    Herpes simplex virus II 109
    varicella-zoster virus 111
    Cytomegalovirus 112
    Epstein Barr Virus 114
    Human herpesvirus–6 115
    Human herpesvirus–7 115
    Human herpesvirus–8 116

Unenveloped DNA viruses

Papovaviridae
    Mouse polyoma virus 118
    SV40 118
    Human polyomavirus
        JC virus 118
        BK virus 118
Human papilloma virus 120
Adenoviridae
    Adenovirus 121
Parvoviridae
    B19 123

RNA Viruses

Enveloped RNA Viruses

Orthomyxovirus
    Influenza 126
        Influenza A virus
        Influenza B virus
        Influenza C virus
Paramyxovirus
    Measles virus 129
    Parainfluenza 129
    Mumps virus 131
    Respiratory syncytial virus (RSV) 131
Togavirus
    Rubella virus 133
    Alpha virus
        Eastern equine encephalitis 134
        Eastern equine encephalitis 134
    Flavivirus
        Saint Louis encephalitis virus 135
        Yellow fever virus 135
        Dengue fever virus 136
    Hepatitis C 138
    Hepatitis E 138
Rhabdovirus
    Rabies virus 139

Table of Contents

The Microbiology Companion, Topf and Faubel ©1997   **V**

Gram + Cocci | Gram – Cocci | Gram + Rods | Gram – Rods | Gram Nothing | DNA Viruses | RNA Viruses | Fungi | Parasites | X–Reference | Index

Table of Contents

The Microbiology Companion, Topf and Faubel ©1997   **VI**

Bunyavirus
  Hantaviruses 140
    Hantaan
    Puumala
    Seoul virus
    Four Corners/Sin Nombre Virus
    Prospect Hill
Filovirus
  Ebola virus 141
  Marburg virus 141
Retroviruses
  The retroviral genes 144
  Human immunodeficiency virus (HIV) 146
    HIV-1
    HIV-2
    Acquired Immunodeficiency Syndrome 147
    prophylaxis 150
  human T-cell leukemia virus 153
    HTLV I
    HTLV II

Unenveloped RNA viruses

Picornavirus
  Enterovirus
    poliovirus pathogen 155
    coxsackievirus 157
    echovirus 157
    hepatitis A (enterovirus 72) 158
  Rhinoviruses 158
Reoviruses
  rotavirus 160
  orbivirus 160

Prions

Prion
Animal Prion diseases
  Bovine Spongiform Encephalopathy 162
  Scrapie 162

Human Prion diseases
  Cruetzfeldt-Jakob Disease 163
  Kuru 163
  Gerstmann-Straussler-Scheinker disease 163
  Fatal familial insomnia 163

Section three: Fungi   165

Yeasts and Molds

Systemic Mycoses
  Coccidioides immitis 167
  Blastomyces dermatitidis 167
  Histoplasma capsulatum 169
  Paracoccidioides brasiliensis 169
Opportunistic Mycoses
  Candida albicans 171
  cryptococcus neoformans 172
  aspergillus fumigatus 173
  Zygomycetes 173
    Mucor
    Rhizopus
Cutaneous Mycoses
  Dermatophytes 175
    Epidermophyton
    Microsporum
    Trichophyton
  malassezia furfur 176
  cladosporium werneckii 176
  sporothrix schenckii 176

Section four: Parasites   178

Protozoa

Ameba
Entamoeba histolytica.............................179

Flagellates
Intestinal Flagellates
Giardia lamblia..............................180
Trichomonas vaginalis.................181
Hemoflagellates
Leishmania
donovani.............................182
braziliensis.........................182
mexicana...........................182
tropica...............................182
Trypanosoma
cruzi..................................184
brucei
brucei..........................185
gambiense....................185
rhodesiense..................185

Sporozoans
Cryptosporidium..........................186
Plasmodium
falciparum........................188
vivax.................................188
ovale.................................188
malariae............................188
toxoplasma gondii........................190
Pneumocystis carinii........................191

Helminths

Platyhelminthes (flatworms)
Cestodes
Taenia saginata.............................196
Taenia solium...............................197
Diphyllobothrium latum................198
Echinococcus granulosus..............199

Trematodes 201
schistosoma
mansoni..............................204
japonicum...........................204
hematobium 204
Clonorchis sinensis....................205
Paragonimus westermani............206

Nematodes (round worms)

Intestinal nematodes
Enterobius vermicularis 209
Trichuris trichiura........................210
Ascaris lumbricoides 211
Hookworms
Ancylostoma duodenale............212
Necator americanus 212
Strongyloides stercoralis..............213
Tissue Nematodes
Trichinella spiralis........................217
Ancylostom braziliense................218
Ancylostoma caninum................218
Toxocara canis 218
Wuchereria bancrofti 219
Dracunculus medinensis..............220
Onchocerca volvulus 221
Brugia malayi 221
Loa loa.......................................222

Section five: Cross-Reference   225

Table of contents for Cross-References

Characteristics
Cause Granulomas.........................226
Capsule...226
Intracellular Existence 226
Motile..226
Hemolysis (beta or alpha hemolytic).....227
Normal Flora.................................227
Catalase positive 228

Table of Contents

Gram + Cocci | Gram – Cocci | Gram + Rods | Gram – Rods | Gram Nothing | DNA Viruses | RNA Viruses | Fungi | Parasites | X–Reference | Index

Table of Contents

IgA protease 228
Recurrent fevers.......................................228
Dimorphic fungi.......................................228
Fungi which produce asexual spores.....................228

Cultures
grown on artificial media or cell cultures 230
grown only in cell cultures...........................230
culture is dangerous..................................230
special culture requirements 230

Toxins
toxins outside of GI tract 231
GI Toxins..232
GI disease without a toxin 232
Diarrhea 233

Transmission
arthropod vectors....................................234
Congenital infections.................................235
Infections acquired in the birth canal..................235

Meningitis
most common cause of meningitis 236
causes of meningitis in special cases.....................237
spinal tap..237

Immunology

Antibodies
IgG 241
IgM 241
IgA 242
IgD 242
IgE 242

Hypersensitivity
Type I anaphylactic....................................243
Type II cytotoxic 243
Type III immune complex disease......................244
Type IV delayed type or cell-mediated..................244

Vaccines
passive immunizations................................246
routine childhood vaccines............................247
vaccination schedule.................................248
active vaccines..249

Glossary **253**

Index..**260**

The first 95% of any job worth doing takes 95% of the time and the last 5% takes the other 95% of the time.

Why

While taking the microbiology course, we, like most of our classmates, spent hours of our study time putting together charts and lists in an attempt to organize the material into a learnable format. While studying for boards we again needed concise tables and lists to digest the information. It was clear that there needed to be a book which organized the information the way students learn.

So, after boards, while our peers were off skittering around the globe, we dusted off our microbiology textbooks and began to process the class into a logical and convenient format. Our goal was to produce the type of document that we would have liked to have had while we were taking micro and studying for boards. Our hope is that our book can help you spend your time learning the material instead of battling with it.

How to use this book

Think of The Companion as a prearranged notebook. Take it with you everywhere, to lectures, lab, the library, the bathroom, wherever you learn micro. When you find or are told an important piece of information about an organism, put it in The Companion. This way you will be able to find it again, instead of having it lost in your notes or the margins of your textbook.

Make lists and keep them in the book. Microbiology tests typically have two types of questions: straight forward and whoa!

The straightforward question gives you the bug and asks for the characteristics:

    Strep. pneumo is: a. G+ b. G- c. acid fast d. a mold

The whoa question gives you the characteristic and asks you for the bug:

    All of the following are normal flora except ...

So, it is important to learn the organisms by characteristics; this requires creating lists. We compiled lists of some the most important features in a special section at the back of The Companion, but there are plenty of blank pages for you to make your own favorite lists (i.e. bacteria which are the hardest to spell; diseases I would least like to get).

Features of the book

One of our goals in writing The Companion was to create an easy quick reference to all of the organisms. This is the bulk of the book. Each medically important pathogen is presented in an easy-to-scan table with all of the important facts on one or two pages. This will save you the frustration of wading through chapters of unintelligible microbabble looking for basic information.

We also tried to provide a framework by categorizing and grouping the pathogens as logically as possible. Each new section outlines the pathogens to be discussed in that section; this provides the reader with a quick summary of the upcoming organisms which helps to make the details more learnable.

Introduction

What's Up With the Revised Second Edition

After we published the second edition years passed and more and new and exciting changes occurred in the filed of infectious disease. Our book started getting out of date and we decided to touch it up. In this newest release we have:

- Up-dated our treatment recommendations
- Expanded coverage of *H. pylori*
- Updated and expanded the HIV/AIDS section
- Added the Hanta virus
- Added the Ebola virus
- Expanded coverage of Human Herpes Viruses
- Added a section on prions

Expanded the section on vaccines

What Was New in the Second Edition

With the input and criticism of many medical students who used the first edition we set out to improve and perfect the book. First off, we went through the book page by page rewriting, clarifying and fine tuning the material. We added charts, rearranged information and improved the flashcards. Then, we added:

- Flashcards
- Wider margins and blank pages so you can add your notes
- Expanded glossary and index
- Spiral binding and more durable laminated covers
- Pretty pictures between each section
- More introductory pages and summary charts

How (and Why) to Contact Us

We believe that this book will help you get through Micro and serve as an excellent review source for boards. We've worked hard to make this book as user-friendly, organized and mistake-free as possible. Please let us know if the book has been useful to you (and feel free to let next year's class know too). If you find any mistakes or have any suggestions on how to make the book better, we'd love to hear that, too. The authors can be reached via email:

Sarah's email address is: SFaublel@yahoo.com
Joel T.'s email address is: JTopf@yahoo.com

If you have any other questions or comments including questions on how to get this book, or ideas for new products, please contact Joel Smith, the president of our little company. He can be reached many ways:

1 (888) ALERT-44 (toll free) (voice and fax)

Alert & Oriented Publishing Co.
7850 El Paseo Grande #5
La Jolla, CA 92037

or via email from our website at WWW.ALERTandONLINE.com

Acknowledgments

Thanks to all the people who bought the first edition and offered their input for the second edition. Other people that we would like to acknowledge are:

Dennis Franklin, Cranbrook Kingswood High School ('95) and future medical student, who created the artwork for the section dividers.

Warren Capell, Resident at University of Colorado who was instrumental in getting files to Joel when Sarah was on-call (writing books as a resident is hard)

Joe Lash, Wayne State Law School ('94), now a practicing attorney, who taught us about intellectual property law.

Jeff Zonder, Wayne State University School of Medicine ('95), now at Rochester for internal medicine, who created the cover in less than a week.

About the Authors

You might be curious "What kind of crazy medical students have the time or energy to write a book?" Good question. Counter-intuitive as it may seem the answer is that they are somewhat organizationally challenged. Sarah lived her entire first year of medical school with a chase lounge as a bed; Joel doesn't have a clue without his Newton. Because of this perhaps they have a better sense then most about what really needs to be organized and how it should be done.

Sarah and Joel authored this book in the summer of 1993, both before and after the "boards", and even into their third year rotations. They have now both graduated from Wayne State University School of Medicine and are residents: Sarah in internal medicine at the University of Colorado and Joel in combined med

icine/pediatrics at Indiana University in Indianapolis. They are currently spending the luxurious amount of free time residency provides writing their second book, a study guide on fluids, electrolytes and acid-base balance.

About Alert and Oriented Publishing

As the Microbiology Companion took shape on Sarah's and Joel's Powerbooks they showed drafts to friends and family. One friend, Joel Smith, avidly encouraged them to sell the book nationwide. One discussion turned to many and eventually the authors agreed to let Smith sell the book at the University of Michigan, where he was beginning a joint J.D./M.B.A. program. The book sold out 4 times in three weeks. After more discussion, what now seems a natural partnership began to form. Joel and Sarah could not deal with the daily barrage of bookstores, sales representatives, printers, and taxes while they were sleepless in surgery but they were able to schedule blocks of time to write a second edition, and begin another book. Smith's enthusiasm (based partly on having been a frustrated biochemistry major) and ever increasing business and law experience made him the perfect partner to take care of the nuts and bolts of running an apartment based publishing company. The three incorporated Alert and Oriented Publishing in the fall of 1993.

The book was written with Microsoft 5.1® on various Apple computers including a Powerbook Duo 2300, a 7500/100 (his), a Powerbook 520c and a 7600/120 (hers). The master copy was printed on a Hewlett Packard 6MP (Jumbo).

Abbreviations

Abbreviations used in the text

1°	primary
2°	secondary
3°	tertiary
AB	antibody
AC	air conditioners
AFB	acid fast bacillus
AGN	acute glomerular nephritis
AIDS	acquired immunity deficiency syndrome
BAP	blood auger plate
BCG	Bacillus Calmette-Guérin (tuberculosis vaccine)
CSF	cerebral spinal fluid
DIC	disseminated intravascular coagulation
DNA	deoxyribonucleic acid
ELISA	enzyme linked immunosorbent assay
ER	endoplasmic reticulum
GBC	general binding company
GI	gastrointestinal tract
HIV	human immunodeficiency virus
IM	intramuscular
INH	isoniazid
IUD	intra uterine device
IV	intravenous
IVDA	Intravenous drug abuse
LPS	lipopolysaccharide
LRI	lower respiratory tract infection
MRSA	Methicillin Resistant Staphylococcus *aureus*
MSSA	Methicillin Sensitive Staphylococcus *aureus*
NCI	National Cancer Institute
PID	pelvic inflammatory disease
RBC	red blood cell
RER	rough endoplasmic reticulum
RES	reticuloendothelia system
RNA	ribonucleic acid
URI	upper respiratory tract infection : sinusitis, epiglottitis, otitis media
UTI	urinary tract infection

Section one: Bacteria

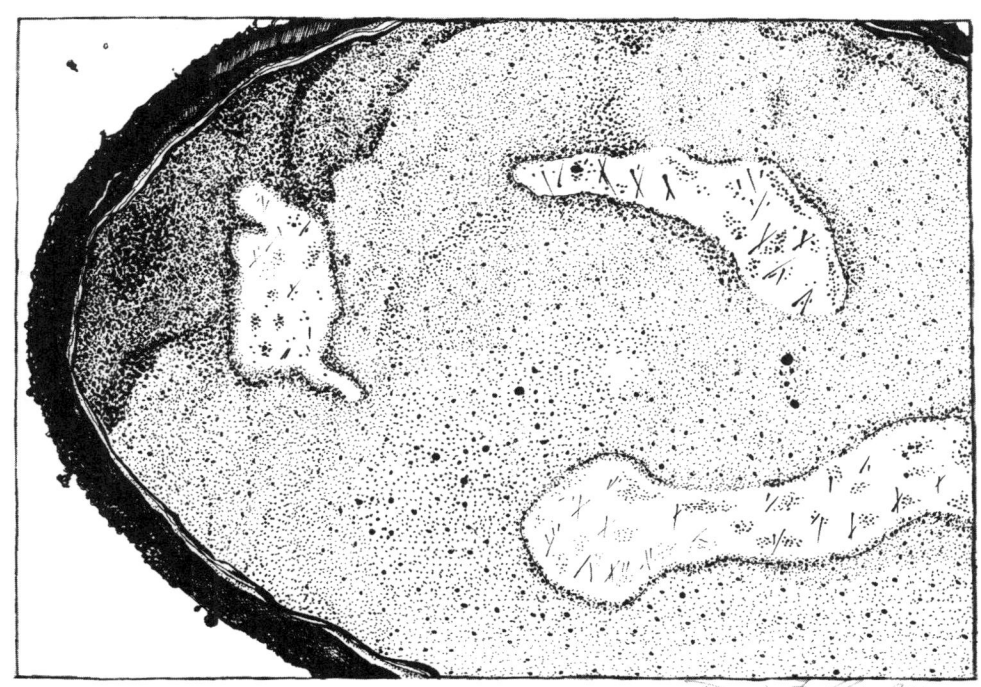

Bacillus fastidiosus, Dennis Franklin

Bacteria

Bacteria are prokaryotic organisms which are smaller and simpler than eukaryotic cells. All animals, plants, molds and parasites are eukaryotic. The numerous differences between prokaryotic and eukaryotic cells have been exploited to create the various antibiotics available today.

feature	eukaryotic	prokaryotic	antibiotics
nuclear membrane	**Present**, consists of a lipid bilayer with pores that restrict which molecules can enter the nucleus.	**Absent**, instead they have a nuclear region where the DNA is associated with the plasma membrane.	
DNA	Associated with **histones**, contains introns.	Not associated with any protein, **lacks introns**.	
chromosomes	**multiple**	**One** chromosome of circular DNA and plasmids. Plasmids are small loops of extrachromosomal DNA which are easily exchanged among bacteria.	Quinolones (ciprofloxacin, norfloxacin, nalidixic acid) inhibit bacterial DNA gyrase which is necessary for DNA replication and RNA production.
translation of RNA	Translation occurs after RNA processing. Translation is never concurrent with transcription.	Translation occurs during transcription of RNA.	Rifampin blocks the ß subunit of the RNA polymerase. Inhibits mRNA synthesis.
ribosomes	**Bigger**: 80S made up of 40S and 60S subunits.	**Smaller**: 70S made up of 30S and 50S subunits.	Aminoglycosides and tetracyclines bind the 30S subunit; chloramphenicol, clindamycin and erythromycin bind the 50S subunit. Inhibit protein synthesis.
cell wall	Some eukaryotic cells have a cell wall made of cellulose. Animal cells do not have a cell wall.	Present except in Mycoplasma and L-forms. Made of peptidoglycan.	ß-lactams (penicillin, cephalosporins) and vancomycin inhibit cell wall synthesis.
organelles	present	none	
folic acid	Require preformed folic acid in the diet.	Make it from the precursor PABA.	Sulfonamides and trimethoprim inhibit folic acid synthesis.

⊃ Pages 150 - 151 of *The Pharmacology Companion* have an even better summary of antibiotic classification and action.

The Bacteria

Gram + Cocci

Bacterial Cocci

The bacterial cocci are divided between gram positive and gram negative cocci.

Genera species	gram stain	diagnostic reactions	diseases
Staphylococcus aureus (coagulase +, beta hemolytic) epidermidis (coagulase -) saprophyticus (coagulase -)	positive	catalase positive	abscesses food poisoning toxic shock syndrome pneumonia
Streptococcus pyogenes (Group A, beta hemolytic) agalactiae (Group B, beta hemolytic) bovis (Group D, usually alpha hemolytic) pneumoniae (alpha hemolytic) Viridans group (alpha hemolytic)	positive	catalase negative	pharyngitis, rheumatic fever pyoderma, acute glomerulonephritis pneumonia meningitis subacute bacterial endocarditis
Enterococcus faecalis faecium	positive	catalase negative	endocarditis urinary tract infections
Neisseria meningitidis (ferments maltose) gonorrhoeae (cannot ferment maltose)	negative	oxidase positive	meningitis urethritis PID

Staphylococcus

- gram + cocci
- facultative anaerobes
- catalase +

These ubiquitous bacteria generally cause abscesses. When viewed under the microscope their colonies form grape-like clusters. The catalase reaction is used to differentiate *Staphylococcus*, which are catalase positive, from *Streptococcus* and *Enterococcus* which are catalase negative.

pathogen	appearance and lab findings	diseases	surface molecules, toxins and enzymes	notes/important properties
Staphylococcus epidermidis	gram + cocci white colonies no hemolysis **catalase +** **coagulase -** **novobiocin sensitive** facultative anaerobes	**Nosocomial infections** The organism frequently infects indwelling catheters and prosthetic joints. **Endocarditis**		Normal flora of skin. Frequent contaminant of specimens obtained through the skin (i.e. blood samples, spinal taps). Resistant to many antibiotics. Treatment may require removal of prosthetic devices.
Staphylococcuss aprophyticus	gram + cocci no hemolysis **catalase +** **coagulase -** **novobiocin resistant** facultative anaerobes	**Urinary tract infections (UTI)** are community acquired and occur particularly in young women (16-30).		Novobiocin resistance is used to differentiate *saprophyticus* from *epidermidis*. Resistant to many antibiotics.

pathogen	appearance and lab findings	diseases	surface molecules, toxins and enzymes	notes/important properties
Staphylococcus aureus	gram + cocci in clusters Colonies have a light golden pigment. **Beta hemolytic**: clear zone around colonies. **catalase +** **coagulase +** facultative anaerobes *Fermentation of Manitol*	**Food poisoning** is characterized by nausea, vomiting, abdominal cramps and watery diarrhea. Symptoms appear within 1-6 hours after ingestion of *pre-formed toxin* (this is a relatively short incubation time). Symptoms usually resolve within 24 hours. **Toxic shock syndrome** (TSS) is a serious acute illness characterized by high fever, erythematous rash, hypotension and shock. May occur in menstruating women using highly absorbent tampons and in men and women with surgical wound infections. **Scalded skin syndrome** is mediated by the exfoliative toxin and is characterized by a diffuse erythematous rash and sloughing of the skin. **Skin infections**: folliculitis boils furuncles carbuncles **Osteomyelitis** (bone infection) *S. aureus* is the most frequent organism causing osteomyelitis. Wound infections Mastitis Endocarditis	**Protein A** is an antiphagocytic cell wall protein which binds the Fc portion of IgG to prevent complement binding and opsonization. **Enterotoxin** is a heat stable toxin which causes the symptoms of food poisoning. **Toxic shock syndrome toxin-1** (TSST-1) is the primary etiological agent of TSS. It is thought to induce the release of TNF and IL-1, which is also the mechanism by which endotoxin induces shock. **Exfoliative toxin** is an exotoxin coded by a plasmid which causes scalded skin syndrome. **Alpha toxin** causes lysis of PMNs and platelets resulting in abscesses. **ß-lactamase** (penicillinase) is a virulence factor produced by most *S. aureus* which confers resistance to penicillin G. The gene is located on a transferable plasmid. **Altered penicillin binding proteins** are virulence factors of a subset of *S. aureus* which confers resistance to all ß-lactam antibiotics.	Coagulase + reaction is used to differentiate *S. aureus* from all other *Staphylococcus* species. **Treatment** usually includes a beta-lactamase (penicillinase) resistant penicillin such as **nafcillin**. *S. aureus* with altered penicillin binding proteins are known as "**methicillin-resistant**" *Staph aureus* or MRSA. MRSA are resistant to all penicillinase resistant antibiotics and the drug of choice is **vancomycin**.

Notes

Gram + Cocci

Streptococcus
- gram + cocci in chains
- facultative anaerobes
- catalase -

Streptococcal infections are often pyogenic (pus forming) and are a special concern for individuals who have had a splenectomy (removal of the spleen). All *Streptococci* have a C carbohydrate in their cell walls, except the Viridans group and *Streptococcus pneumoniae*. Streptococci are arranged into **groups** (Lancefield Groups A-T) based on antigenic differences in the C carbohydrate. The principle groups are A, B and D.

pathogen	appearance and lab findings	diseases	surface molecules, toxins and enzymes	notes/important properties
Streptococcus pyogenes Group A Strep	gram + cocci in chains **Beta hemolytic**: clear zone around colonies. Lancefield Group A **bacitracin sensitive** facultative anaerobe Identification is by antibody agglutination of group A specific antigens. Anti Streptolysin O (ASO) titers rise following infection.	**Pharyngitis** is a self limited infection of the pharynx characterized by sore throat, headache, fever, and malaise. Classically, the pharynx is erythematous and a grey-white purulent exudate is visible on enlarged tonsils. **Transmission** is by respiratory droplets. Most common type of group A infection. **Scarlet Fever** occurs in patients with *pharyngitis* caused by a strain of *S. pyogenes* that produces erythrogenic toxin. Characteristics include a generalized erythematous rash which feels like sandpaper and a strawberry-appearing tongue. The organism remains in the pharynx but the toxin becomes disseminated.	**Pili** are virulence factors which allow the organism to adhere to respiratory mucosa. **M protein** is an antiphagocytic surface molecule which induces type specific immunity. There are over 70 antigenic types. **Streptokinase** promotes lysis of blood clots by converting plasminogen to plasmin. Is used to treat myocardial infarcts. **Streptolysin O** is an oxygen labile hemolysin which induces antibodies. **Streptolysin S** causes beta hemolysis on a BAP.	All group A Strep are susceptible to penicillin G. *S. pyogenes* is the only *Streptococcus* which is bacitracin sensitive. 5-30% of the population are asymptomatic carriers. *S. pyogenes* is the most frequent bacterial cause of pharyngitis. Overall, viruses are the most frequent cause of pharyngitis. **Erythrogenic toxin** causes scarlet fever. The toxin is produced by a lysogenic temperate bacteriophage which is carried by some (but not all) strains of *S. pyogenes*. ⟶ mechanism similar to TSST of S. Aureus. *acts as superantigen*

pathogen	appearance lab findings	diseases	notes/important properties
Streptococcus pyogenes Group A Strep continued		**Pyoderma** (impetigo) is a highly contagious skin infection characterized by a rash which forms **crusty yellow sores** over the entire body. Particularly common among children. Transmission is by contact with sores and contaminated objects such as towels. Associated with poor personal hygiene. grp A **Non-suppurative sequelae:** • **Acute rheumatic fever** (ARF) is a systemic disease characterized by inflammatory lesions which may appear in the heart, joints, subcutaneous tissues and central nervous system. May follow any group A infection, usually pharyngitis. May be prevented by treating pharyngitis with penicillin within 9 days of onset. The disease is the result of damage caused by antistreptococcal antibodies which cross react with antigens in the joints and heart. This is an example of a **type II hypersensitivity reaction**. • **Acute glomerulonephritis** (AGN) is the prototype of acute nephritic syndrome. The disease is characterized by: 　1. **cola colored urine** (2° to hematuria, blood in the urine) 　2. **hypertension** 　3. **edema** The disease is a result of damage caused by the accumulation of antigen-antibody complexes on the glomerular basement membrane. This is an example of a **type III hypersensitivity reaction**. AGN usually follows Group A infection by 7-21 days and is most common in children 3-10 years of age. Under age 6, AGN follows pyoderma but in older children and young adults, it more commonly follows pharyngitis.	C carbohydrate – determines the group of B hemolytic strep. located in cell wall M protein – most imp. virulence factor – determines type of B hemolytic strep. Interferes w/ ingestion of phagocytes. Provides types specific immunity – Most strep part of normal flora of human throat, skin, intestines but produce disease when gain access to tissues of blood

Gram + Cocci

pathogen	appearance and lab findings	diseases	cell surface molecules, toxins and enzymes	notes/important properties
Streptococcus agalactiae Group B Strep (GBS) Think: Group B for OB (obstetrics)	gram + cocci **Beta hemolytic**: clear zone around colonies. Lancefield Group B facultative anaerobe **bacitracin resistant** Identification is by antibody agglutination of group B specific antigens.	**Neonatal infections** are separated into two clinical groups: early and late onset. **Early onset** typically appears within 48 hours and appears as **pneumonia** and **sepsis**. This serious infection carries a 50-60% mortality rate and is the result of maternal-infant transmission during delivery. GBS is the most frequent cause of neonatal sepsis in the U.S. **Late onset** usually appears as **meningitis**. It occurs during days 10-60 and is a less severe infection. The source of the organism is nosocomial (from the hospital nursery). Postpartum endometritis Chorioamnionitis		Normal flora of the GI tract. Group B Strep are part of the **normal vaginal flora** in 5-35% of all women. Pregnant women are often screened for GBS at about 28 weeks. **Pregnant women** who are GBS+ are given **amoxicillin** during labor to prevent transmission to the infant. Combined antibacterial treatment with **penicillin** and an **aminoglycoside** is needed for neonatal infections.

pathogen	lab findings	diseases	toxins and enzymes	notes
Streptococcus pneumoniae (pneumococcus)	gram + diplococci lancet shaped **Alpha hemolytic**: green zone around colonies. **bile soluble** **inhibited by optochin** **+ quellung reaction** facultative anaerobe No C carbohydrate so it is not classified by a Lancefield group.	**Lobar (typical) pneumonia** is characterized by a sudden onset of fever, cough, pleuritic pain and rusty sputum. Bacteremia can occur. 90% of all lobar pneumonia is caused by *S. pneumoniae.* **Meningitis** *S. pneumonia* is one of the most common causes of bacterial meningitis in adults and children. Otitis media Sinusitis	**Capsule** is antiphagocytic. Antibodies to the capsule provide type specific immunity. There are 85 antigenic types. **IgA protease** inactivates mucosal IgA which favors colonization.	**Pneumococcal vaccine** contains 23 types of capsular antigens which represent 90% of all infections; recommended for people greater than 65 years old, splenectomized immunocompromised, or have sickle cell disease. Treat with penicillin G or erythromycin. Some penicillin resistance is emerging. **bacitracin resistant**
Viridans group	gram + cocci **Alpha-hemolytic**: green zone around colonies. **not bile soluble** **not inhibited by optochin** facultative anaerobe No C carbohydrate so it is not classified by a Lancefield group.	**Subacute bacterial endocarditis** May occur after dental work in people with previously damaged heart valves. Antibiotic prophylaxis is recommended for this population. Causes emboli, murmurs and anemia. **Dental carries** is caused by *S. mutans* which produces dextrans, a component of plaque.		Normal flora of **oropharynx**. Bile insolubility and resistance to optochin differentiate Viridans from pneumococcus; both organisms are alpha hemolytic. Treat with penicillin G. **bacitracin resistant**

pathogen	appearance and lab findings	diseases	cell surface molecules, toxins and enzymes	notes/important properties
Streptococcus bovis (Group D Strep)	gram + cocci Usually **alpha hemolytic**: green zone around colonies. Lancefield Group D facultative anaerobe no growth in 6.5% NaCl	**Urinary tract infections** **Subacute bacterial endocarditis** (see below)		**bacitracin resistant**
Enterococcus faecalis faecium	gram + cocci Usually **alpha hemolytic**: green zone around colonies. Lancefield Group D **grows in 6.5% NaCl**	**Secondary infection** in urinary tract or wounds. **Subacute bacterial endocarditis** (see above) Meningitis		Normal flora of the intestines. Treat with penicillin G, ampicillin or vancomycin; add gentamycin if meningitis or endocarditis is present.

Gram Positive Cocci, Group D Strep and *Enterococcus*

Gram + Cocci

- Diagnosis depends on clinical Findings, cerebral spinal Fluid chemistry hematology + microbiology

- Gram stain of cerebral spinal Fluid can detect bacteria in 50% of cases of purulent meningitis + can distinguish b/t H. inflenze, Strep pneum, N. mening

Neisseria

- ◆ gram negative
- ◆ kidney shaped diplococci
- ◆ **oxidase positive**

Neisseria are grown on chocolate agar and Thayer-Martin medium. All *Neisseria* have an **IgA protease** which cleaves mucosal IgA and facilitates colonization of mucosal surfaces. Humans are the only host.

pathogen	appearance and lab findings	diseases	surface molecules, toxins and enzymes	notes/important properties
Neisseria meningitidis (meningococcus)	gram - diplococci **oxidase positive** **+ quellung reaction** ferments maltose Identified by latex agglutination test for capsular antigens in spinal fluid. Classified into serotypes A, B, C, D, X, Y, Z, 29E and W-125 on the basis of capsular antigens. Grows on chocolate agar and Thayer-Martin medium. *Vancomycin - inhibits gram(+)* *colistin - inhibits gram(-)* *nyastin - inhibits yeast* *capnophilic*	**Meningitis** infection begins with colonization of the upper respiratory tract and is transmitted via respiratory droplets. The illness presents with a maculopapular rash and fever which have a simultaneous onset. **Epidemics** can be caused by serotypes A, B, and C. Most epidemics have been associated with group A. Outbreaks are common among individuals living in close quarters, i.e. **military personnel.** **Meningococcemia** may cause Waterhouse-Friderichsen syndrome (WFS) which is characterized by fever, shock, DIC and massive hemorrhage into the adrenal glands which results in their destruction. This is a cause of <u>primary adrenal insufficiency</u> (Addison's disease).	**Capsule** is antiphagocytic. **Endotoxin** may cause WFS and DIC. IgA protease *endotoxemia - thromboembolic*	A **vaccine** containing capsular antigens is available which protects against A, C, Y, and W-135 variants. Given to people at risk (military recruits) or during epidemics. Serotype B, a cause of epidemics, cannot be incorporated into a vaccine because it is poorly immunogenic in humans. Use **rifampin** for *or ciprofloxacin* prophylaxis in individuals who have close contact with infected persons. Treat with penicillin G or ceftriaxone.

petechial rash = hallmark

Colonies exposed the phenylenediamine turn black as a result of oxidation of reagent by the enzyme

pathogen	lab findings	diseases	surface molecules, toxins and enzymes	notes
Neisseria gonorrhoeae (gonococcus)	gram - diplococci oxidase positive (possess enz. cytochome c) cannot ferment maltose Indirect immunofluorescence test of patient's serum can detect antibodies to gonococcal antigens. Grows on chocolate agar and Thayer-Martin medium.	**Gonorrhea** is a reportable STD with different symptoms in men and women. **Males**: urethritis, purulent discharge and pain during urination. **Females**: infection is usually localized to the endocervix causing a purulent vaginal discharge. May progress to pelvic inflammatory disease (PID) and sterility. **Gonococcal arthritis** occurs as a result of disseminated infection and affects the joints, heart and skin. Individuals who are deficient in late acting complement components (C5-C8) are at a higher risk for a disseminated infection. **Ophthalmia neonatorum** is an eye infection acquired during passage through the birth canal of an infected mother. Infection is prevented by use of erythromycin ointment or silver nitrate in neonates' eyes.	**Pili** are the primary virulence factor (gonococci lacking pili do not cause disease). They inhibit phagocytosis and allow attachment to host cells. Vary antigenically so that antibodies are not protective. **Membrane Protein II** is a surface protein which is also involved in attachment to the host. IgA protease	2nd most common STD after chlamydia. A major problem with controlling gonorrhea is the existence of asymptomatic carriers who spread the disease and do not seek treatment. Infection is often asymptomatic in women. Treat with a one time IM injection of ceftriaxone. Treatment must also include an agent active against chlamydia (i.e. doxycycline) because it is a common co-pathogen.

Handwritten notes: oxidase (+) gram (-) enough to justify treatment in uncomplicated cases. No serologic tests. ♂ - gram (-) intracellular diplococ on direct smear from urethral exudate = diagnostic. can also use urine specimen. recommended for sites other than anogenital. Definitive Dx = sugar utilization tests or slide agglutination. not routine. Auxotyping - characterizes strains based on nutritional requirements. AHU Auxotype (req. arginine, hypoxanthine, uracil) v. susceptible to penicillin - resistant to bactericidal activity of normal human sera + complement. Piliated = virulent, Non piliated = avirulent. OMP variation due to regulation @ translation. cell wall blebs which slough off peptidoglycan + endotoxin. can survive in PMN's & non-pro phagocytes.

- ELISA
- Genetic transformation tests to detect gonococcal DNA
- DNA hybridiz. test
Vulvovaginitis - ♀ 2-8yr, sexual abuse, 10% asymptomatic ♂, 40% " ♀

Gram Negative Cocci, Neisseria   Gram - Cocci   The Microbiology Companion, Topf and Faubel ©1997   **13**

Gram Positive Rods

The gram positive rods are composed of four genera:

spore formers

Bacillus — **only aerobic gram + rod that forms spores**

anthracis — anthrax

cereus — food poisoning

Clostridium — **only anaerobic gram + rod that forms spores**

botulinum — food poisoning, infant botulism

tetani — wound infections

perfringens — gas gangrene

difficile — pseudomembranous colitis

non-spore formers

Corynebacterium

diphtheriae — diphtheria

Listeria — **does not produce exotoxins**

monocytogenes — opportunistic meningitis

Actinomyces israelii is also a gram positive rod but it will be discussed with the other Actinomycetes.

Gram + Rods

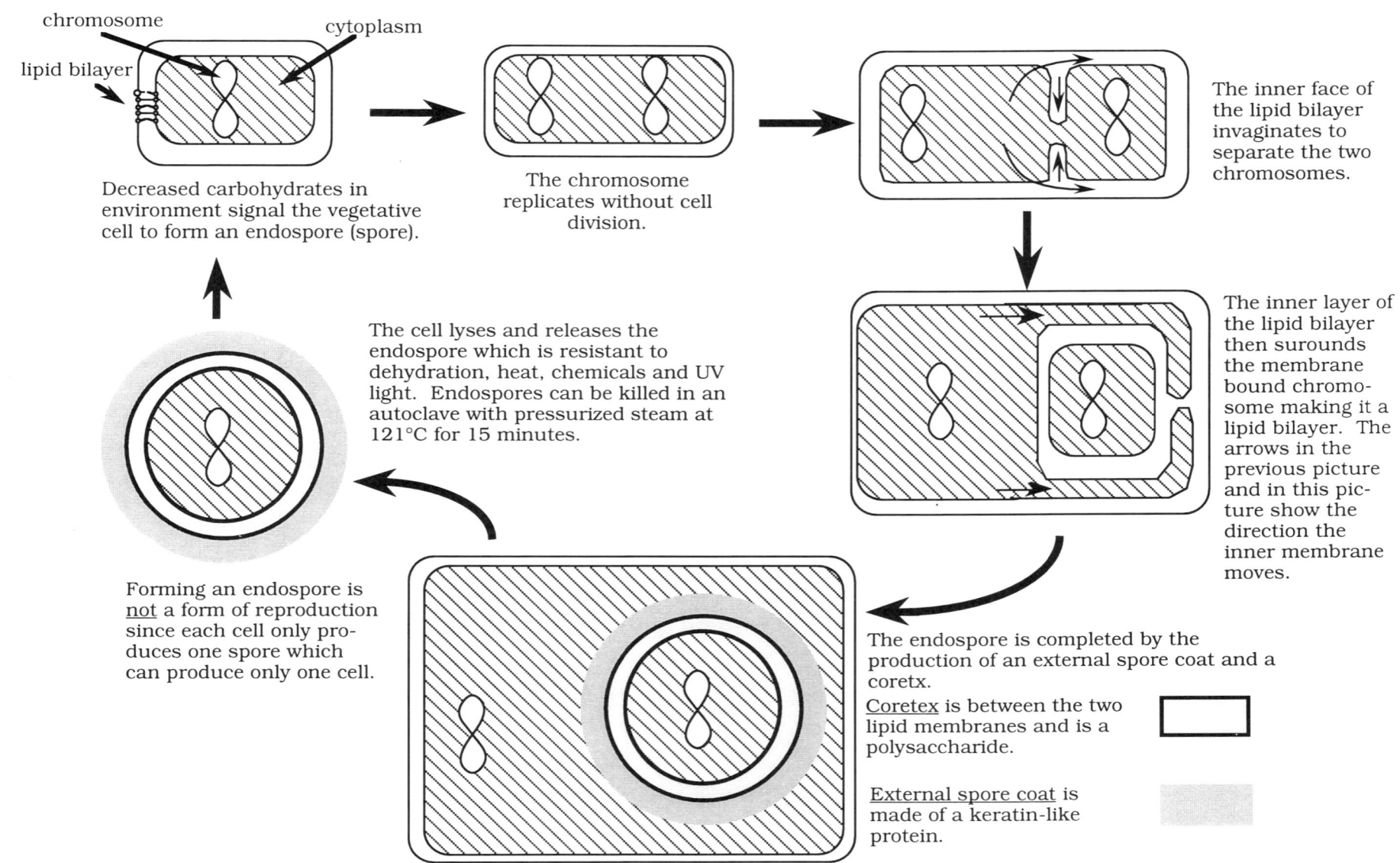

chromosome

cytoplasm

lipid bilayer

Decreased carbohydrates in environment signal the vegetative cell to form an endospore (spore).

The chromosome replicates without cell division.

The inner face of the lipid bilayer invaginates to separate the two chromosomes.

The cell lyses and releases the endospore which is resistant to dehydration, heat, chemicals and UV light. Endospores can be killed in an autoclave with pressurized steam at 121°C for 15 minutes.

The inner layer of the lipid bilayer then surounds the membrane bound chromosome making it a lipid bilayer. The arrows in the previous picture and in this picture show the direction the inner membrane moves.

Forming an endospore is <u>not</u> a form of reproduction since each cell only produces one spore which can produce only one cell.

The endospore is completed by the production of an external spore coat and a coretx.

<u>Coretex</u> is between the two lipid membranes and is a polysaccharide.

<u>External spore coat</u> is made of a keratin-like protein.

pathogen	lab findings	diseases	toxins and enzymes	notes
Bacillus anthracis	gram + rod colonies grow in long chains produces **spores** catalase + nonmotile Facultative anaerobe which **requires O$_2$ to produce spores**.	**Anthrax** is primarily a disease of animals which causes death in 2-3 days. Spores from infected animals remain in the soil for years. Humans get two types of anthrax: **Cutaneous** disease is the most common type. **Transmission** is by contact with infected animal hides. A black necrotic lesion forms on the skin with profuse local edema. If left untreated can cause bacteremia and death. **Pulmonary** disease (**wool sorters disease**) is more serious than the cutaneous form. **Transmission** is by inhalation of spores.	**Anthrax toxin** is a complex toxin with 3 components: **Edema factor** (EF) is an adenylate cyclase ($\uparrow$cAMP) which interferes with the phagocytotic activity of PMNs. Requires PA for entry into cells. **Lethal factor** (LF) **Protective antigen** (PA) binds to host cells and allows entry of EF. EF and PA act analogously with the A and B subunits of other toxins. The major difference being that EF and PA are separate exotoxins not subunits of the same exotoxin. **Capsule** is anti-phagocytic and has two unique characteristics: • Composed of **protein** rather than polysaccharides. • Contains the amino acid **D-glutamate** which is unusual because *all* other natural proteins are composed of L-isomers.	**Vaccine** contains protective antigen (PA) which is available for persons at risk such as cowboys, cowgirls and wool sorters. Treat with ciprofloxacin or doxycycline. Separate **plasmids** code for the capsule and anthrax toxin.
Bacillus cereus	gram + rod produces **spores** **aerobe** motile	**Food poisoning,** two types: Short incubation (1-6 hr) • caused by the ST toxin • nausea and vomiting Long incubation (10-12 hr) • caused by the LT toxin • abdominal pain and diarrhea **Reheated fried rice** is a common source of poisonings.	**Two enterotoxins**: **Heat stable** (ST) causes vomiting. **Heat labile** (LT) increases the production of cAMP similar to the *E. coli* LT toxin.	The symptoms of food poisoning are caused by **preformed toxins**, therefore, antibiotic therapy is useless.

Spore Forming Gram Positive Rods, *Bacillus*

Gram + Rods

Gram + Rods

Clostridia
- **gram + rods**
- **spore forming**

Each species of *Clostridia* produces **exotoxins** which are responsible for disease. *Clostridia* are the only spore forming gram + rods which are **obligate anaerobes**. The 4 species are described below.

Clostridium	toxin	diseases	
botulinum	**Botulinum toxin** Blocks the release of acetylcholine at the neuromuscular junction.	type	where spores germinate
		Food borne botulism	in a can
		Infant botulism	in the gut
		Wound botulism	in a wound
		Each form of botulism produces the same symptoms: diplopia, weakness, flaccid paralysis.	
tetani	**Tetanospasmin** (tetanus toxin) Inhibits the release of inhibitory neuro-transmitters (GABA, glycine). Vaccine contains tetanus toxoid.	**Tetanus** (lockjaw) Symptoms: violent muscle contractions, spastic paralysis.	
perfringens	**Enterotoxin** HL, causes food poisoning. **Exotoxins** Cause tissue necrosis.	**Food poisoning** Occurs after ingestion of spores. Watery diarrhea. **Gas gangrene** Occurs when spores germinate in a wound. Gas is produced in the wound.	
difficile	**Exotoxins** Toxin A: diarrhea. Toxin B: cytotoxin.	**Pseudomembranous colitis** Necrosing inflammatory lesion of the colon. Treat with metronidazole (Flagyl) or oral vancomycin.	

Gram + Rods

pathogen	diseases	toxin	notes
Clostridium botulinum gram + rod spore forming obligate anaerobe motile	*C. botulinum* causes three clinical forms of botulism. All three are mediated by the botulinum toxin and result in clinically similar disease. **Food borne botulism** **Transmission** occurs by the **ingestion** of **preformed toxin** contained in food. *C. botulinum* spores from the environment contaminate meats and vegetables (the bacteria cannot grow on fruit which is too acidic). After being canned, the spores germinate in the anaerobic conditions of the can and the organism produces toxin. Spores are resistant to heat; sterilization requires 121°C moist heat for 15 minutes. Some cases result from home canners who neglect this precaution. Incubation period of 1-4 days. Nausea, vomiting and diarrhea are the earliest symptoms which are followed by the classic neurological symptoms. **Infant botulism** **Transmission** occurs by the **ingestion** of **spores** contained in contaminated food (**honey** in particular). Unlike adults, infants lack a normal colonic flora which suppresses the growth of *C. botulinum.* The spores germinate in the intestine and produce toxin. Occurs in infants 3 to 35 weeks. May account for some cases of Sudden Infant Death Syndrome (SIDS). **Wound botulism** **Transmission** occurs when **spores** enter a wound, usually traumatic. In the wound, the spores germinate and produce toxin. The toxin disseminates throughout the body and eventually reaches nerve terminals to cause the symptoms of botulism.	**Botulinum toxin** is a polypeptide which **blocks the release of acetylcholine** from cholinergic neurons. Most symptoms of botulism are due to the decrease of acetylcholine at the <u>neuromuscular junction</u> which causes: • **diplopia**, earliest neurologic sign • dysphagia • progressive weakness • **flaccid paralysis** • death from **respiratory failure** *C. botulinum* is divided into 8 serotypes based on the type of toxin it produces: A, B, C alpha, C beta, D, E, F and G. Types **A**, **B** and **E** are responsible for the majority of human botulism. The lethal dose for humans is about 1 microgram (10^{-6}g), one of the most potent toxins known. Encoded by a lysogenic phage.	**Home canned foods** are the most frequent source of botulism in the U.S. Cans that are swollen should be tossed. *C. botulinum* produces gas which swells cans. Although the spores are resistant to heat, the toxin is heat labile and can be inactivated by several minutes of boiling. **Treatment** includes A, B, and E antitoxin and respiratory support. Because the disease is mediated by a toxin, antibiotic therapy is useless.

Notes

Gram + Rods

Spore Forming Gram Positive Rods, *Clostridia*

pathogen	diseases	cell surface molecules, toxins and enzymes	important properties
Clostridium tetani gram + rod spore forming obligate anaerobe drumstick appearance motile	**Tetanus** (lockjaw) **Transmission** occurs when a wound is infected with **spores** which germinate and produce toxin. Ischemic tissue (i.e. anaerobic environment) facilitates germination of spores. **Tetanus toxin** (tetanospasmin) travels in peripheral neurons to the CNS where it **inhibits the release of inhibitory neurotransmitters** (i.e. glycine and GABA) and causes: • violent muscle contractions • **spastic paralysis** • death from **respiratory failure**	**Tetanospasmin** (tetanus toxin) is produced as a single protein which is cleaved into two polypeptide chains linked by a disulfide bond: **H chain** 　larger (100,000 daltons) 　binds toxin to neural tissue **L chain** 　smaller (50,000 daltons) 　exerts biological effect of toxin Encoded by a **plasmid**. Organisms without this plasmid are not virulent.	Spores are widespread and found in soil and feces. **Vaccine** contains **tetanus toxoid** and is routinely administered to infants (DPT).

Tetanus Immunization

Immunization history is important when treating traumatic injuries. The proper dosing schedule is as follows:

DPT (diphtheria and tetanus toxoids with pertussis vaccine) should be given at:

2 months	4 months	6 months	15-18 months	4-6 years	total of five doses

DPT should only be given to children *under* 7 years of age.

Td (tetanus toxoid with a reduced dose of diphtheria toxoid) should be given at:

14-16 years	plus	every 10 years throughout life.

Td should only be given to persons *over* 7 years of age.

Tetanus immunization history	Clean, minor wounds <6 hours old, < 1 cm in depth, no dirt or contaminants in the wound.	All others wounds contaminated with soil, saliva or feces; puncture, crush or frost bite wounds; >6 hour old, > 1 cm deep
Uncertain or incomplete (less than three doses)	Td only	Td and TIG (given at separate sites)
Complete (three or more doses) and up to date	nothing	nothing
Complete, but greater than 5 years and less than 10 years since last dose.	nothing	Td only
Complete, but greater than 10 years since last dose	Td only	Td only

- ◆ **Td** is the adult vaccination.
- ◆ **TIG** is tetanus immune globulin which provides immediate passive immunity to the tetanus toxin.
- ◆ When both Td and TIG are given, they are injected at separate sites. Otherwise, the immune globulin would bind to the toxoid and both would be rendered useless.

Gram + Rods

pathogen	diseases	cell surface molecules, toxins and enzymes	important properties
Clostridium perfringens ◆ gram + rod ◆ spore forming ◆ obligate anaerobe ◆ nonmotile	**Food poisoning** **Transmission** occurs when **spores** are ingested and germinate within the small intestine. The incubation period is 8-16 hours. Infection is characterized by a watery diarrhea which resolves within 24 hours. **Gas gangrene** **Transmission** occurs when wounds are infected with **spores** which germinate in the anaerobic environment of tissue necrosis. *Clostridium perfringens* ferments carbohydrates and produces large quantities of gas. The gas causes high pressures which can limit blood supply and enhance the anaerobic environment needed by the pathogen.	**Enterotoxin** is heat labile and causes food poisoning. **Exotoxins** cause extensive **tissue necrosis** around an infection: • **alpha toxin (lecithinase)** damages cell membranes • **mu toxin (hyaluronidase)** lyses RBCs, cardiotoxic • **iota toxin** is responsible for **acute enterotoxemia** *Clostridium perfringens* secretes a total of 12 different types of exotoxins.	Normal flora of colon. Spores are widespread and found in soil and feces. Antibiotic treatment of food poisoning is not necessary. The primary treatment of gas gangrene is surgical removal of infected and dead tissue. Antibiotic therapy is with penicillin G plus clindamycin.
Clostridium difficile ◆ gram + rod ◆ spore forming ◆ obligate anaerobe ◆ motile	**Pseudomembranous colitis** is a necrotizing inflammatory lesion of the colon. **Diarrhea** is the chief symptom. *C. difficile* is part of the normal flora of 3% of the population. Because *C. difficile* is resistant to most antibiotics, when persons carrying the organism are treated with broad spectrum antibiotics (**clindamycin** and ampicillin, are classic causes but the most common cause are the cephalosporins). *C. difficile* can overgrow, produce toxins and cause disease.	Two **exotoxins**: **Toxin A** Enterotoxin which causes diarrhea. **Toxin B** Cytotoxin which kills cells and requires toxin A for activity.	Treat with metronidazole or oral **vancomycin.** Antibiotic induced pseudomembranous colitis is the only indication for *oral* vancomycin, which is usually given IV 2° to poor intestinal absorption. Since little of oral vancomycin is absorbed large amounts reach the colon to treat the infection.

Notes

Gram + Rods

Corynebacterium

- non-spore forming
- gram + rod

pathogen	appearance and lab findings	diseases	cell surface molecules, toxins and enzymes	important properties

Corynebacterium diphtheriae	gram + rods
	
	pleomorphic
	
	non-spore forming
	
	facultative anaerobe
	
	Cytoplasm has **granules** which contain phosphate energy stores.
	
	Metachromatic staining: when stained with methylene blue or toluidine blue the granules stain red.
	
	Two selective media allow isolation:
	
	   Loeffler medium
	
	   BAP with potassium tellurite
		Diphtheria is an upper respiratory tract infection.
		
		Transmission occurs by respiratory droplets.
		
		The organism remains localized in the throat and produces exotoxin which enters the blood and causes damage in the kidneys, heart and nerves.
		
		A gray **pseudomembrane** forms in the back of the throat due to accumulation of necrotic tissue. This may lead to obstruction of the airway and death.
		
		The incubation period is 2 to 6 days.
			Diphtheria toxin inhibits protein synthesis in all eukaryotic cells by enzymatically attaching ADP to elongation factor 2 (EF-2). This inactivates EF-2 which is required to move the ribosome down the mRNA to read the next codon.
			
			Coded for by a **lysogenic phage** that is incorporated into the bacterial DNA.
			
			The toxin is produced as a single protein which is cleaved into two polypeptide chains.
			
			  peptide A is the active enzyme
			
			  peptide B binds to cells and is required to get the A peptide into the cell
			
			A good way to remember the mechanism of this toxin is: <u>C D E F</u>, just like the alphabet:
			
			<u>C</u>orynebacterium <u>D</u>iphtheriae blocks <u>E</u>longation <u>F</u>actor.
				Patients with diphtheria are given IV erythromycin, but the toxin which causes the morbidity is not inactivated by antibiotics.
				
				Treatment and prevention both depend on neutralizing the toxin.
				
				If the clinical picture indicates diphtheria an **antitoxin** is given. The disease is so dangerous that there is no time to wait for confirmatory labs.
				
				Vaccine contains diphtheria toxoid which is routinely administered to infants (DPT).
				
				Diphtheria is very **uncommon** in the U.S., less than 10 cases per year.

Listeria

- ◆ non-spore forming
- ◆ gram + rod

pathogen	appearance and lab findings	diseases	toxins and enzymes	important properties
Listeria monocytogenes	gram + rod **Beta hemolytic**: narrow clear zone around colonies. **tumble** motility facultative anaerobe **LPS-like component** in cell wall (experimentally produces symptoms like gram negative endotoxin). **Intracellular reproduction** in mononuclear phagocytes. catalase +	Immunocompromised, neonates and pregnant women are at particular risk for infection. **Transmission** of infection varies. The organism is primarily a disease of animals, so contact with animals or animal products (milk) serves as a source of transmission. The organism is also widespread in the environment, and transmission may result from the ingestion of contaminated vegetables. Finally, neonates may be infected in utero when the organism crosses the placenta. **Meningitis** is the most common manifestation of infection with *L. monocytogenes*. Leading cause of bacterial meningitis in cancer patients and **renal transplant recipients.** **Neonatal infections** occur in two forms: **Early onset** occurs on the first or second day of life and causes **pneumonia**. The infant is critically ill. Acquired **in utero** or during delivery. **Late onset** occurs at 1-4 weeks and causes **meningitis.** Acquired during the delivery. **T- cell immunity** is protective. Infection may cause **granulomas.**	**Hemolysin** is the only known virulence factor. It is similar to Streptolysin O.	Infection can occur from contact with soil, animal feces, contaminated vegetables, and especially **milk**. Pasteurization does not entirely destroy the pathogen. One outbreak occurred from a Mexican type of cheese. Treat with ampicillin.

Gram + Rods

Gram + Rods

Notes

Gram + Rods

The Microbiology Companion, Topf and Faubel ©1997

Gram Negative Rods are divided into the following categories: enterics, anaerobes, respiratory infections and zoonotics.

Enterics

Enterobacteriaceae
- *Escherichia coli*
- *Shigella*
- *Salmonella*
- *Klebsiella pneumonia*
- *Enterobacter cloacae*
- *Serratia marcescens*
- *Proteus*
- *Providencia*
- *Morganella*
- *Yersinia*
  - *pestis*
  - *enterocolitica*

Pseudomonadaceae
- *Pseudomonas aeruginosa*

Vibrionaceae
- *Vibrio*
  - *cholera*
  - *parahaemolyticus*

Campylobacteraceae
- *Campylobacter jejuni*
- *Helicobacter pylori*

Anaerobes

- *Bacteroides*
  - *fragilis*
- *Prevotella*
  - *melaninogenica*
- *Fusobacterium*

Respiratory Infections

- *Haemophilus*
  - *influenza*
  - *aegyptius*
  - *ducreyi*
- *Legionella pneumophila*
- *Bordetella pertussis*

Zoonotic Bacilli

- *Brucella*
  - *abortus*
  - *melitensis*
  - *suis*
- *Francisella tularensis*
- *Yersinia*
  - *pestis*
  - *enterocolitica*

Summary of the Enteric Gram Negative Bacilli

Enteric bacilli are gram negative rods which are either normal inhabitants or pathogens of the digestive tract. The enterics consist of four families:

Family	Oxidase	Grows on MacConkeys	metabolism of glucose	typical diseases
Enterobacteriaceae	-	+	ferments glucose	see below
Pseudomonadaceae	+	+	oxidizes glucose	pneumonia, UTI
Vibrionaceae	+	+	ferments glucose	cholera, diarrhea
Campylobacteraceae	+	-	inert to glucose	diarrhea, gastritis

Gram – Rods

Enteric bacilli consist of four families and each family is made up of multiple genera and species:

organism	characteristics/toxins/enzymes	diseases
ENTEROBACTERIACEAE *Escherichia coli*	Capsule Flagella O antigen Pili	Diarrhea UTI Meningitis Gram - septicemia

Enterotoxigenic
ETEC
	Heat labile toxin
	   B subunit: binds cell membranes
	   A subunit: ↑**cAMP**
	Heat stable toxin: ↑ **cGMP**
		Traveler's diarrhea
		   Voluminous diarrhea without blood or mucous.
		   Primary site of infection is the small intestine.

Enteroinvasive EIEC	Ability to invade intestinal epithelial cells.	**Dysentery** Fever, abdominal cramps with blood and pus in the stool.
Enteropathogenic EPEC	EPEC adhesion factor Unable to invade intestinal epithelium.	Diarrhea and vomiting with mucous in the stool.
Enterohemorrhagic EHEC	**Shiga-like toxin** • One A subunit inactivates 60S ribosome and stops protein synthesis. • Five B subunits bind glycolipid on intestinal mucosa.	**Hemolytic colitis** Abdominal cramps, bloody diarrhea without pus.

Shigella
  dysenteriae
  flexneri
  boydii
  sonnei
	Shiga toxin
	• One A subunit inactivates 60S ribosome and stops protein synthesis.
	• Five B subunits bind glycolipid on intestinal mucosa.
		Dysentery
		   Bloody mucopurulent diarrhea.
		   Transmission by fingers, flies, food and feces.

Salmonella
  typhi
  cholera suis
  paratyphi
	Capsule
	Flagella
	Intracellular replication in monocytes.
		Enterocolitis (gastroenteritis)
		   Invades intestinal mucosa.
		   Bacteremia is uncommon.
		Typhoid fever
		   Infection begins in the intestines.
		   Bacteremia is common.
		   Rose spots often present on the abdomen.
		   Organism infects the gallbladder.

organism	characteristics/toxins/enzymes	diseases
Klebsiella pneumoniae	Urease + Capsule Coliform	Opportunistic and nosocomial infections Pneumonia
Enterobacter cloacae	Motile Coliform	Opportunistic and nosocomial infections Urinary tract infections
Serratia marcescens	Coliform	Opportunistic and nosocomial infections Pneumonia
Proteus *Providencia* *Morganella*	Urease + (except *Providencia*) Very motile Coliform	Urinary tract infections

Yersinia
  pestis
  enterocolitica
	anti-phagocytic capsule
	intracellular existence
	Heat stable toxin: ↑ **cGMP**
		Bubonic Plague: transmission is by flea bites. Respiratory transmission can occur, pneumonic plague.
		Enterocolitis identical to Shigella, arthritis can be a late complication.

PSEUDOMONADACEAE *Pseudomonas aeruginosa*	Oxidase + Motile Produces blue and green pigments. Exotoxin A: inactivates EF-2.	**Opportunistic infections**: pneumonia, burn infections **Nosocomial infections**: UTI, meningitis. **Swimmer's ear**

VIBRIONACEAE
Vibrio cholera
	Oxidase +
	Flagellum
	Enterotoxin (choleragen)
	  5B subunits: bind ganglioside GM-1
	  1A subunit: ↑**cAMP**
		Cholera
		  Rice water stools.
		  Fecal oral transmission.

Vibrio parahaemolyticus	Grows on 8% NaCl	**Food poisoning**
		  Nausea, vomiting and diarrhea.
		  Transmission by ingestion of seafood.

CAMPYLOBACTERACEAE
Campylobacter jejuni
	Enterotoxin
	  Heat labile: ↑**cAMP**
		Enterocolitis
		  Abdominal pain, bloody diarrhea.

Helicobacter pylori	Urease +	**Chronic gastritis** **Duodenal ulcers**

Gram Negative Rods, Enteric Bacilli

Gram – Rods

Enterobacteriaceae family has many genera, some are compared in the following chart.

If an organism is able to ferment lactose it will appear as a red colony on MacConkeys agar.

Genera	ferment lactose	produce H$_2$S gas	motility	produce indole	produce urease	typical diseases
Escherichia	yes	no	yes	**yes**	no	UTI, diarrhea, meningitis
Shigella	**no** except *S. sonnei*	no	**no**	no	no	dysentery
Salmonella	**no**	**yes**	yes	no	no	typhoid fever, enterocolitis
Klebsiella	yes	no	**no**	no	**yes**	pneumonia, UTI
Enterobacter	yes	no	yes	no	no	pneumonia, UTI
Serratia	slow fermenter	no	yes	no	no	pneumonia, UTI
Proteus	no	**yes**	yes swarms	**yes**	**yes**	UTI

Remember *all* Enterobacteriaceae share these traits:

◆ oxidase -
◆ grow on MacConkeys agar
◆ ferment glucose

Notes

Gram – Rods

Enterobacteriaceae, *E. coli*

pathogen	lab findings	diseases	toxins and enzymes
Escherichia coli	gram - rod flat green colonies on EMB agar red colonies on MacConkeys agar does not produce H$_2$S gas oxidase - ferments glucose lactose fermenter facultative anaerobe phenylalanine deaminase + (produces indole)	**Diarrhea** The *E. coli* that cause diarrhea produce virulence factors not found in *E. coli* of normal flora. The diarrhea producing *E. coli* are further categorized into: enterotoxigenic, enterohemorrhagic, enteroinvasive and enteropathogenic strains, which are described in the table below. **Urinary tract infections** *E. coli* is the most frequent cause of UTI. Infection can be limited to the bladder, **cystitis**, causing dysuria (painful urination) and frequency, or the infection can ascend into the kidney, **pyelonephritis**, causing fever and flank pain. **Systemic infections** **Meningitis**: *E. coli* is the second most frequent cause of meningitis in neonates. **Gram - septicemia**: *E. coli* is the most frequent cause of nosocomial gram - septicemia.	**Capsule** antigens are serotyped and designated by K. **Flagella** antigens are serotyped and designated by H. 40 serotypes. **O antigen** of the endotoxin is a lipopolysaccharide. 150 serotypes. **Pili** are virulence factors which allow attachment to mucosal surfaces. Bind to mannose. **Enterotoxins** are toxins encoded by plasmids which are produced by various strains of *E. coli* and cause diarrhea. These toxins are further discussed below. **Normal flora** of the colon.

E. coli strain	disease	enterotoxin and its mechanism
Enterotoxigenic ETEC	**Traveler's diarrhea** (Turista) is characterized by voluminous diarrhea without blood or mucus. Short duration, 1-3 days. The organism does not invade the intestinal mucosa. Disease is mediated by toxins. Primary site of infection is in the **small intestine**. Rare in developed countries.	**Heat labile toxin** (LT) has a mechanism similar to cholera toxin. It is plasmid encoded and has 2 subunits: **B subunit**: binds to the cell membrane. **A subunit**: permanently stimulates adenylate cyclase which ↑**cAMP**. cAMP signals the mucosa to **secrete Cl⁻** into the intestinal lumen and **water follows** passively. Large amounts of water and electrolytes can be lost from the small intestine. **Heat stable toxin** (ST) is plasmid encoded and stimulates guanosine cyclase which ↑ **cGMP** and prevents the transport of ions out of the lumen. **Colonizing factor antigen** is a plasmid encoded **pilus** required for infection.
Enteroinvasive EIEC	**Dysentery** is characterized by fever, abdominal cramps and blood and pus in the stool. Clinically identical to the dysentery caused by *Shigella*.	Primary virulence factor is the ability to **invade the epithelial cells of the intestinal mucosa**. Does not become systemic. Ability to invade cells is encoded by a plasmid.
Enteropathogenic EPEC	Diarrhea and vomiting. Stools contain mucus. Primary site of infection is in the **small intestine**.	**EPEC adhesion factor** is coded by a plasmid and allows the organism to adhere to the intestinal mucosa and causes destruction of microvilli. Unable to invade intestinal epithelium.
Enterohemorrhagic EHEC	**Hemolytic colitis** is characterized by severe abdominal cramps and large amounts of **bloody diarrhea** without pus. **Hamburger meat** is a frequent source of infection. **Hemolytic-uremic syndrome** is characterized by hemolytic anemia, decreased platelets, and acute renal failure. **O157:H7** is the most common serotype.	**Shiga-like toxin I** and **II** (SLT I, II) **each have six subunits**, one A subunit and five B subunits. These toxins are antigenically and functionally similar to *Shigella* toxins. The **B subunit** (5 copies) binds a specific glycolipid on the intestinal mucosa. The **A subunit** (1 copy) inactivates the 60S ribosome and stops protein synthesis. This **kills the intestinal lining cells** and prevents resorption of fluid. This results in bloody diarrhea.

Gram – Rods

pathogen	appearance	diseases	cell surface molecules
Shigella *dysenteriae* *flexneri* *boydii* *sonnei*	gram - rod nonmotile does not produce H$_2$S gas oxidase - grows on MacConkeys agar ferments glucose **Does not ferment lactose** (except *S. sonnei* which is a slow lactose fermenter). **Positive methylene blue** stain of fecal sample indicates the presence of PMNs. (All invasive infections of the GI: EIEC, *Shigella*, *Salmonella* and *Campylobacter*, show PMNs in a fecal sample.)	**Dysentery** is characterized by bloody mucopurulent diarrhea which occurs after an incubation of 1-4 days. Invades intestinal mucosa producing local inflammation (PMNs) and ulceration. **Does not cause bacteremia.** Transmission is by the Four Fs: **Fingers, Flies, Food** and **Feces** (fecal-oral route is common among toddlers). ***S. dysenteriae*** causes the most severe disease. ***S. sonnei*** causes mild dysentery and accounts for the majority of *Shigella* infection in the U.S.	**Shiga toxin** has six subunits **B subunit** (5 copies) binds glycolipid on the intestinal membrane. **A subunit** (one copy) inactivates the 60S ribosome and stops protein synthesis. This kills the cells lining the intestine and prevents resorption of fluid. All species produce this toxin in varying amounts. *S. dysenteriae* produces 1000 times the Shiga toxin as *S. sonnei*.

- invasiveness of colon is main virulence factor

a) Genes encoding invasion antigens are on large plasmid
b) Invasion involves actin polymerization to propel organisms to move laterally for cell-cell spread

pathogen	appearance	diseases	cell surface molecules
Salmonella typhi *cholerae suis* *paratyphi* *Salmonella* is like a fish (salmon) which swims around the body (systemic infection) while *Shigella* is like a shingle that just covers the GI mucosa (non-systemic infection).	gram - rod does not ferment lactose motile facultative anaerobe produces H$_2$S gas oxidase - grows on MacConkeys agar ferments glucose **Produces gas** during glucose fermentation (except *S. typhi*). **Positive methylene blue** stain of fecal sample indicating the presence of PMNs. **Intracellular:** able to **multiply in monocytes** and avoid humoral immunity. Each species has hundreds of serotypes based on O antigens; the serotypes are named for where they were isolated.	**Transmission**: *Salmonella* infection is most commonly transmitted via ingestion of contaminated **poultry and eggs**; contaminated water and contact with infected animals are other sources. **Enterocolitis** (gastroenteritis) The organism invades intestinal mucosa and the underlying tissue, producing local inflammation, with PMNs, and diarrhea after an incubation period of 10-24 hours. **Bacteremia is uncommon**. The dose required to produce infection is very high: 100,000 organisms. The low pH of the stomach is an important host defense. **Typhoid fever** is an enteric fever caused by *S. typhi* after an incubation period of 1-3 weeks. Infection begins in the intestine (without symptoms). Invasion of local lymph nodes leads to spread of the organism throughout the body. **Bacteremia is common**. **Rose spots** appear when organism multiplies in the skin, often present on the abdomen. When the organism infects the **gallbladder**, a carrier state may result with an asymptomatic host shedding organism into stools.	**Capsule**: a virulence factor identified as antigen Vi. **Flagella**: exists in two different forms, H1 and H2, only one of which is expressed at a given time. This allows the bacteria to vary its antigenic appearance and evade antibody immune response. The switch between flagella is controlled by the H2 promoter. See "phase variation" in the glossary. Toxins for *Salmonella* have yet to be determined.

Enterobacteriaceae, *Salmonella*

Gram – Rods

Salmonella → 3 distinct clinical syndromes:
　　　　　1) gastroenteritis
　　　　　2) enteric fever
　　　　　3) septicemic syndrome

A) Salmonella gastroenteritis (food poisoning)

Disease: Cramping, vomiting, diarrhea (rotten egg smell) headache. Fever. Blood invasion can occur. Self limiting

Organisms: Any of the 1700 sp. of Salmonella except typhi. Most common causative organism in US = S. typhimurium

Transmission: Zoonosis (animal → man) via food from infected animals

Pathogenesis: Surface adhesions mediate attachment + invasion of mucosal cells of small int. (Shigella invades large intestine) A toxin may mediate fluid excretion

Diagnosis: Fecal culture processed on selective & diff. agar. Biochem tests. Slide agglutinations for serogroup ID if necessary. Salmonellae classified in grps A→E on basis of O antigen. W/in a group species is diff by presence of another O antigen + flagellar + Vi antigens

Immunity: no vaccine

Treatment: Flouroquinones, Ciprofloxacin

B) Enteric Fevers

Disease: initially malaise, headache, anorexia
1st wk: Gradual ↑ in fever 7-10 days after onset of symptoms. Constip. or diar. Blood (+)
2nd wk: Rose spots in 50% cases (7-14 days) Constant high fever, (+) stool cult. 14 days get septicemia
3rd wk: Fever remits. Gradually organisms shed back to bowel. Can have ulceration, bloody stool

Organisms: S. typhi → caus. agent of typhoid fever
　　　　S. paratyphi, S. schotmülleri cause milder types

Pathogenesis: actually a systemic disease. organs ingested & multiply in s.i. At ileum, penetrate gastric mucosa thru M cells at peyers patches → lymph → blood. Bacteremia of 3-4 days. cleared by MØ of liver & spleen → bacteria multiply here. Gall bladder may be infected + gain reentry to intestine

Transmission: only affects man. thru direct contact or contaminated food or H₂O

Diagnosis: Isolation via agar & serotyping. Organisms most freq found in blood in 1st wk. 2nd wk → stool

c) Septicemic Syndrome

Organism: S. choleraesuis

Disease: Spiking fevers. intest. symptoms may be absent. Freq. seeding to various organs → abscesses, meningy. pneumonia, osteomyelitis

therapy: Short course Flouroquinones

immunity: Anti-Vi + anti-O antib. Active inf. → cellular immunity → effective → infect activates MØ
1) killed whole cells → toxicity
2) live, attenuated, oral vaccine
3) Vi injected capsular polysacc. vaccine

pathogen	lab findings	diseases	toxins and enzymes	notes
Klebsiella pneumonia	gram - rod **+ quellung reaction** glucose and lactose fermenter nonmotile urease + does not produce H_2S gas	Most commonly associated with **opportunistic** and/or **nosocomial** infections. **Lobar pneumonia** is characterized by the production of a thick bloody, "currant jelly" sputum. Urinary tract infections Burn wound infections	**Capsule** is large and inhibits phagocytosis. Over 70 antigenic types. Anti-capsular antibodies provide type specific immunity. Coded on a plasmid.	**Coliform.** Found in 5-10% of healthy individuals. Frequently has multiple antibiotic resistance. Treat with an aminoglycoside and a third generation cephalosporin until sensitivity results are known.
Enterobacter cloacae	gram - rod **- quellung reaction** glucose and lactose fermenter **motile**	Most commonly associated with **opportunistic** and/or **nosocomial** infections. **Urinary tract infections**		**coliform** Frequently resistant to ampicillin and cephalosporins.
Serratia marcescens	gram - rod **slow lactose fermenter** ferments glucose **motile** does not produce H_2S gas	Most commonly associated with **opportunistic** and/or **nosocomial** infections. **Pneumonia**		**coliform**
Proteus Providencia Morganella	gram - rod nonlactose fermenter **urease +** (except *Providencia*) **phenylalanine deaminase +** (produces indole) only *Proteus* produces H_2S gas **Very motile**: do not form colonies but <u>swarm</u> over the agar.	**Urinary tract infections** are almost as common as those caused by *E coli*. These can be either nosocomial or community acquired.	**Urease** converts urea to NH_3 and CO_2 which increases the pH of urine. Increased urinary pH (more alkaline) has two effects: • Predisposes to **renal stone** formation. • Permits colonization of bacteria causing urinary tract infections.	**coliform** Found in feces, sewage and soil. Cell wall O antigens of *Proteus* induce antibodies which cross react with some species of *Rickettsia*. This is the basis of the Weil-Felix reaction in *Rickettsia*.

The genera *Yersinia* is a member of the Enterobacteriaceae but it will be listed with the arthropod borne infections.

Enterobacteriaceae

Gram – Rods

Gram – Rods

Enteric bacilli, family: **Pseudomonadaceae**

Pneumonia   Sepsis   endocarditis   urinary tract inf

- ◆ oxidase +
- ◆ grows on MacConkeys agar
- ◆ oxidizes glucose
- ◆ only one medically important genera and species

→ due to structure of outer membrane porins which restrict entry

pathogen	appearance and lab findings	diseases	cell surface molecules, toxins and enzymes	notes/important properties
Pseudomonas aeruginosa	gram - rod obligate aerobe does not ferment lactose oxidase + motile produces two pigments: • **pyocyanin** is **blue** • **pyoverdin** is **green** and fluorescent can grow in **tap water** *Fruity-grape odor*	**Opportunistic infections** occur in immunocompromised hosts, especially people with **cystic fibrosis**, **burns** and cancer. Pneumonia *diabetics* Eye infections, particularly in contact lens wearers. Skin, burn, and wound infections **Nosocomial infections** **Urinary tract infections** from indwelling urinary catheters. **Meningitis** from inoculation during a lumbar puncture. **Pneumonia** is often severe. Predisposing factors include prolonged ICU stay, antibiotic use and lung disease. **Swimmer's ear**, (otitis externa) is an infection of the external ear canal.	**Exotoxin A** inhibits protein synthesis by the same mechanism as diphtheria toxin: inactivation of elongation factor 2 (EF-2). Since EF-2 is required for ribosome function, protein synthesis stops and the cell dies. elastase collagenase lecithinase *(phospholipase)* *pili + non pili adhesions*	**Resistant to most antibiotics**. Patients on antibiotic therapy are particularly prone to *P. aeruginosa* superinfections. Treat serious infections with **two antibiotics** (e.g. aminoglycoside plus antipseudomonal penicillin) to ensure adequate treatment and prevent resistance. Frequently contaminates respirators, humidifiers and other nonsterile water sources.

no person → person spread

Gram – Rods

Enteric bacilli, family: **Vibrionaceae**
◆ oxidase +
◆ grows on MacConkeys agar
◆ ferments glucose

pathogen	lab findings	diseases	toxins and enzymes
Vibrio cholera	gram - rod comma shaped **slow lactose fermenter** oxidase + does not produce H2S gas single flagellum Facultative **anaerobe** that grows better in oxygen. No growth on 2% NaCl. The low pH of the stomach is protective, so a large infecting dose is needed. Grows well at high pH (9), persons without gastric acidity are at an increased risk of infection.	**Cholera** is a serious disease characterized by the sudden onset of voluminous, non-bloody, watery diarrhea containing flakes of mucus. The term "**rice water stools**" is used to describe the appearance of the diarrhea. **Transmission** is **fecal-oral** by ingestion of fecally contaminated water or food. The organism **does not become systemic.** The organism does not penetrate intestinal epithelium, but adheres to the gut mucosa. Disease is mediated by the toxin choleragen. Can be **quickly fatal** (within 8 hours) due to the rapid loss of fluid and electrolytes. The loss of fluids results in dehydration and hypotension. Electrolyte imbalance results in cardiac and renal failure. **Treat** with rapid fluid and electrolyte replacement.	**Enterotoxin (choleragen)** is similar to the heat labile toxin LT of *E. coli*. It has 6 subunits: **five B subunits**: bind to ganglioside GM-1. **one A subunit**: permanently stimulates adenylate cyclase which ↑**cAMP**. cAMP signals the mucosa to **secrete Cl⁻** into the intestinal lumen and **water follows** passively. Large amounts of K^+ and bicarbonate are also lost. **Mucinase** is needed for pathogenesis because it enables the pathogen to break down and adhere to the protective mucous barrier of the intestines. Strains of *V. cholera* are divided into two major groups based on the O antigen of the cell wall: **O-1 serotype** is associated with epidemic cholera. **Non-O serotype** is associated with sporadic cholera.
Vibrio para-haemolyticus	gram - rod Same as *V. cholera* except: **Grows on 8% NaCl** Halophilic	**Food poisoning** is characterized by nausea, vomiting and diarrhea. **Transmission** is by ingestion of raw or undercooked **seafood** (particularly shellfish) contaminated with *V. parahaemolyticus*. Common in Japan.	Toxin is similar to choleragen.

Notes

Gram – Rods

Enteric bacilli, family: **Campylobacteraceae**

◆ oxidase +
◆ no growth on MacConkeys agar
◆ inert to glucose

pathogen	appearance	diseases	toxins and enzymes	notes
Campylobacter jejuni	gram- rod curved or S-shaped **urease -** oxidase + single flagellum motile microaerophilic inert to glucose nalidixic acid sensitive Unable to grow at 25°C, must be cultured at 45°. Cannot be grown on MacConkeys agar.	**Enterocolitis** is characterized by abdominal pain severe enough to mimic acute appendicitis. The diarrhea is smelly and usually contains **blood**. **Transmission** (person-to-person) is by the **fecal-oral** route. Animals are the reservoir and persons may also be infected by ingesting raw or undercooked meat, fecally contaminated water and <u>unpasteurized milk</u>.	**Enterotoxin** is heat labile; activates adenylate cyclase and ↑cAMP.	Normal gut flora of animals. Because the infection is **self-limited**, antibiotic therapy is questionable.

pathogen	appearance	diseases	toxins and enzymes	notes
Helicobacter pylori (formerly known as *Campylobacter pylori*)	gram - rod curved or S-shaped flagella motile microaerophilic **urease +** oxidase + inert to glucose Cannot be grown on MacConkeys agar.	**Duodenal** and **gastric ulcers** are associated with *H. pylori* infection of gastric mucosa. **Transmission** is most likely by the fecal-oral route. Eradication of *H. pylori* greatly reduces ulcer recurrence. **Diagnosis** is by biopsy, serology and/or the urea breath test. **Gastric cancer** occurs more frequently in patients with *H. pylori* infection.	**Urease** promotes the enzymatic conversion of urea to ammonia and CO_2. Ammonia raises gastric pH which protects the organism from destruction by the normally acid gastric environment. In the urea breath test, the patient is fed isotopically labeled urea and the breath is monitored for isotopically labeled CO_2.	**Infection increases with age**; a general rule of thumb is that the % infected is the same as the age (i.e. 30% of 30 year olds and 60 % of 60 year olds are infected). **Treatment** of *H. pylori* associated ulcers is with **metronidazole**, **tetracycline**, **omeprazole** (a proton pump inhibitor) *and* **bismuth** (the active ingredient in Pepto-Bismol™) for one week. Other combinations of antibiotics and acid blockers are also used, but the regimen listed above currently has the highest eradication rate (> 95%).

Gram – Rods

Anaerobes

The Microbiology Companion, Topf and Faubel ©1997 **48**

Anaerobic bacilli

pathogen	appearance and lab findings	diseases	cell surface molecules, toxins and enzymes	important properties
Bacteroides fragilis	gram - rod **obligate anaerobe** Grows on BAP with vancomycin and kanamycin (to inhibit other organisms). bile resistant	**Abscesses** of the <u>pelvic</u> and <u>abdominal</u> cavities are endogenous infections which occur as a result of trauma (i.e. surgery, IUDs). *B. fragilis* is in the normal flora of the vagina and gut. Anaerobic infections are predisposed by any factor which reduces available oxygen: loss of blood supply, tissue necrosis, or infection with facultative bacteria (i.e. *E. coli*) which will use up the available oxygen. Infection is often mixed with other bacteria.	**Capsule** is the primary virulence factor. One of the rare gram negative bacteria **without endotoxin activity.**	The species *Bacteroides* is the most common bacteria in the gut, numbering 10^{11} per gram of feces. Resistant to penicillin, tetracycline, cephalosporin. aminoglycoside, and vancomycin. **Treat** with metronidazole or clindamycin.
Prevotella melaninogenica (formerly known as *Bacteroides melaninogenica*)	gram - rod **obligate anaerobe** Produces black pigment when grown on BAP bile sensitive	**Lung abscesses** occur as a result of aspiration of normal oral flora. **Periodontal disease** is contributed to by *p. melaninogenica*. At one time, infection by this organism was a major cause of maternal death during labor.	Capsule	Normal flora of the mouth, vagina and large intestine.
Fusobacterium	gram - rod obligate anaerobe Long slender rods, **tapered** at both ends.	Sinus infections Ear infections Dental infections Brain abscesses Lung infections Often mixed with *Bacteroides*.		Normal flora of the mouth, colon and vagina.

Notes

Gram – Rods

Respiratory Infections
Gram negative rods which are primarily involved in respiratory infections include the genera *Haemophilus*, *Legionella*, and *Bordetella*. Don't forget that *Klebsiella*, *Serratia*, *Pseudomonas* and *Bacteroides* also cause respiratory infections but were discussed earlier.

pathogen	appearance and lab findings	diseases	cell surface molecules, toxins and enzymes	important properties
Haemophilus influenza *Haemophilus influenza* does not cause influenza. Influenza is an upper respiratory infection caused by the influenza virus. *Haemophilus influenza* is a common cause of secondary bacterial pneumonia in viral influenza and this led to the erroneous conclusion that it was the etiologic agent.	gram - coccobacillus Growth requires: **factor X: heme** and **factor V: NAD** X and V are found in chocolate agar, which is made by heating a BAP for 15 minutes at 80°C. *- CO₂ incubation for some strains* *- Gram stain spinal fluid*	All infections with *H. influenza* begin by colonization of the upper respiratory tract; transmission is person-to-person by respiratory droplets. **Meningitis**: *H. influenza* is the leading cause of this serious infection in children aged 6 months to 6 years. High mortality if untreated. **Epiglottitis** is a rare, serious illness which occurs in children. The epiglottis quickly swells and obstructs the airway. **Pneumonia** **Upper respiratory infections** otitis media — *due to non-* sinusitis *encapsulated* bronchitis *form*	**Capsule** is antiphagocytic. 6 serotypes designated **a-f** are based on the capsular antigens. **Type b** causes the most serious disease and is most frequently implicated in meningitis and epiglottitis. Antibodies to the capsular antigens provide type specific immunity. **Vaccine** contains type b capsular polysaccharide conjugated to a carrier protein and is routinely administered to infants. Disease is commonly caused by **nonencapsulated** (nontypeable) strains. **IgA protease** inactivates secretory IgA to allow colonization of the respiratory mucosa.	Multiple antibiotic resistance. Treat serious illness with **ceftriaxone**, a third generation cephalosporin. **Rifampin** is used for prophylaxis in persons who have close contact with infected individuals, particularly children in daycare during meningitis epidemics. *- Amp resis due to plasmid w/ amp res. transposon Found in enteric bacilli*

Think: when her child gets the <u>flu</u>, mom goes to the five (<u>V</u>) and dime (<u>X</u>) to buy <u>chocolate</u> (agar).

In 50% of cases Gram stain can distinguish, H. influenza, S. pneumonia, N. meningitidis of purulent meningitis

pathogen	appearance and lab findings	diseases	cell surface molecules, toxins and enzymes	important properties
Haemophilus aegyptius This species of *Haemophilus* <u>does not cause a respiratory tract infection.</u>	gram - coccobacillus similar to *H. influenza*	**Conjunctivitis**, better known as **pink eye**, is inflammation of the conjunctiva characterized by redness and a profuse mucopurulent discharge.		Treat with topical antibiotic ointment.
Haemophilus ducreyi This species of *Haemophilus* <u>does not cause a respiratory tract infection.</u>	gram - coccobacillus similar to *H. influenza*	**Chancroid** is a <u>sexually transmitted disease</u> characterized by **soft, painful** ulcers in the genital area. Local lymphadenopathy is common.		Treat with one dose of IM ceftriaxone. Primary syphilis is characterized by a hard, painless chancre.

Think: when you get chancroid you do cry (*ducreyi*). Painful lesion.

Gram – Rods

pathogen	lab findings	diseases	toxins and enzymes
Legionella pneumophila	gram - rod stains very lightly Lung biopsies are stained with Dieterle silver impregnation. Culture medium must be enriched with **iron** and **cysteine**. Because the organism is difficult to isolate by culture (needs special staining and culture media described above), **diagnosis** relies on serologic tests demonstrating a rise in antibody titer.	**Legionnaires' disease** is an **atypical pneumonia** named after the outbreak which occurred at the 1976 American Legion convention in Philadelphia. • **Airborne transmission** through contaminated water. • **Air conditioners** are the cause of most outbreaks. • No person to person transmission occurs. The organism may be found in lakes, ponds and tap water. The disease can be fatal. Risk factors for infection include: smoking, alcohol consumption and immunosuppression. **Pontiac fever** is a milder form of infection named after an outbreak in Michigan in 1968. Symptoms include fever, headache, diarrhea and vomiting. Self-limited within 3-4 days.	**Beta-lactamase** confers resistance to penicillin. **Intracellular replication** within macrophages. Treat with erythromycin.
Bordetella pertussis	gram - rod + quellung reaction Grows on Bordet-Gengou medium. ↓ *- 10-15% blood in potato starch base /or 10% charcoal to neutralize inhib sub* *-also low [cephalosporin] or other antib w/ gram⊕ spectra to inhibit normal throat flora* *-slide agglut or fluoresc antib*	**Whooping cough** starts with a runny nose, coughing and sneezing. **Paroxysmal phase** begins 1-2 weeks later and is characterized by violent spasms of coughing which are followed by an inspiratory "whoop." The disease is spread by aerosol droplets before the paroxysmal stage, usually before the diagnoses. Only humans are infected. **Vaccine** contains <u>killed organisms</u> and is routinely administered to infants in combination with tetanus and diphtheria vaccines (DPT). Controversy over the vaccine exists because of post-vaccination reactions, which although rare, do occur. The most serious of which is encephalopathy which has caused death or permanent neurologic damage in some vaccine recipients.	**Capsule** is essential for virulence. *→B subunit = binding A subunit = action* **Pertussis toxin** catalyzes the irreversible activation of G proteins (e.g. adenylate cyclase, phospholipase C). It is a two subunit toxin with the A unit containing the toxin and the B unit binding to the host cell. *- histamine sens, ↑insulin synth, ↑lymphocyte prod, ↓phagocytosis* **Adenylate cyclase** is secreted by the bacteria and inhibits the bactericidal action of neutrophils. *impair H₂O₂ + superoxide* **Binding protein** on **pili** allows attachment to the cilia of the respiratory tract.

use for 1° + booster → 3 DTaP vaccines now (acellular) → inactivated pert. toxin bact. comp. pertussis   Filam. Hemag + pili

booster → DTaP w/ Hib used to booster D,T,P + H. influenza

- Antibiotic of little value during whooping stage

↳ Filamentous hemagglutinin

Tracheal cytotoxin - kills ciliated epith

Notes

Gram – Rod

Zoonotic bacteria cause infections whose transmission is associated with an animal.

pathogen	appearance and lab findings	diseases	cell surface molecules, toxins and enzymes
Brucella *abortus* *melitensis* *suis*	gram - rod aerobic Grows on enriched culture medium. Diagnosed by a rise in antibody titer. A skin test can indicate past exposure (**type IV hypersensitivity**).	**Brucellosis** (undulating fever) is characterized by initial symptoms of fever, chills and drenching sweats. **Transmission**: the disease is an infection of animals, which is transmitted to humans as a result of <u>occupational contact</u> (i.e. slaughterhouse workers) or <u>ingestion of unpasteurized milk</u> products. The organisms localize in the RES and multiply within macrophages. Periodic release of the organisms is associated with bacteremia and recurrent fever. **Granulomas** form throughout the body. Lymphadenopathy, splenomegaly and/or hepatomegaly are common. The disease often relapses months later and may take a chronic course of years.	**Capsule** is necessary for virulence. **Intracellular** growth within macrophages. --- **Control of the disease** **Vaccine** containing live attenuated organism **is given to all cattle**. The vaccine is not for people. Cows that are infected are destroyed. **Pasteurization** of milk is used to control brucellosis. **Treatment** is difficult because the organism hides in macrophages and evades antibiotic therapy. Therapy is usually prolonged with multiple antibiotics.

pathogen	appearance and lab findings	diseases	cell surface molecules, toxins and enzymes
Francisella tularensis	gram - rod facultative **anaerobe** Organism is not cultured because of danger to lab workers. Diagnosed by serology. **Ticks** are the vector and **wild animals** are the reservoir.	**Tularemia** is a **tick borne infection** which affects people who handle wild animals. Skinning rabbits is a common source of infection. **Transmission** is by injection of the organism by the bite of an infected tick. The skin at the site of the bite ulcerates and the organism spreads to local lymph nodes which become sore and tender. From the lymph nodes, the organism spreads to other organs resulting in fever, headache and lesions in the liver, spleen and lungs. Caseating **granulomas** may form throughout the body. The organism is intracellular which allows it to evade humoral immunity and antibiotics. Therefore, relapses are common.	**Intracellular** growth **in monocytes and PMNs** allows it to evade antibodies and complement. **Vaccine** is available to persons at high risk (workers who have contact with the furs of wild animals). The vaccine contains **live attenuated** organisms (the only other live bacterial vaccine used in humans is the BCG vaccine for TB). Treat with streptomycin. **Transovarian** transmission of the organism occurs in ticks.

Gram – Rods

pathogen	appearance and lab findings	diseases	cell surface molecules, toxins and enzymes
Yersinia pestis *Yersinia* is in the family Enterobacteriaceae.	gram - rod The ends of the rod stain darker than the middle so it looks like a safety pin. *Yersinia* is a member of the family Enterobacteriaceae so it is **oxidase negative**, grows on **MacConkeys agar** and **ferments glucose**. **Fleas** are the vector and **rodents** are the reservoir. non-motile	**Plague** (syn. Black Death, Bubonic Plague) The disease has two reservoirs: wild rodents (country rats) which are resistant to infection and urban rodents (city rats) which are susceptible to the infection. When the city rat comes in contact with the country rat, the city rat becomes infected with the disease. As the urban rodent population dies off their fleas become desperate for hosts and begin to bite humans which leads to human epidemics. After the host is bitten by an infected flea, the organism goes to regional lymph nodes which become swollen and painful (buboes). The disease causes subcutaneous hemorrhages and DIC. Lung involvement is possible and pneumonia allows respiratory spread of the disease (pneumonic bubonic plague).	**Capsule** is designated serologically as **Fraction 1** (F1) and is anti-phagocytic. Antibodies to F1 are protective. **Murine toxin** causes irreversible shock and death in rodents. **Intracellular** within macrophages. Treatment is with streptomycin. **Vaccine** containing inactivated bacilli is available for persons at risk, i.e. agricultural workers.
Yersinia enterocolitica Infection by *Y. enterocolitica* is not zoonotic, but causes GI disease as do the enteric gram - rods.	gram - rod motile lactose negative	**Enterocolitis** is characterized by fever, diarrhea, and abdominal pain. It is clinically identical to the enterocolitis caused by *Salmonella* or *Shigella*. **Mesenteric adenitis** may cause abdominal pain severe enough to mimic acute appendicitis. **Transmission** is by ingestion of food contaminated with animal (i.e. dog, cat or cattle) feces. Spontaneous recovery occurs within a few days. **Arthritis** is a late complication which occurs two weeks after the GI symptoms. Septicemia is a rare but highly fatal complication.	**Heat stable toxin (ST)** works in the same manner as the *E. coli* ST toxin and increases cGMP in the cell. This prevents the transport of ions out of the lumen and across the mucosa. Antibiotic treatment for enterocolitis and mesenteric adenitis is not necessary; septicemia is treated with an aminoglycoside.

Gram – Rods

Spirochetes

Spirochetes are a family of bacteria whose cell walls are coiled and flexible. The cell wall is **gram negative-like** in that it contains an outer membrane, a peptidoglycan layer and an inner cytoplasmic membrane. They are highly **motile** with a **flagella-like** structure, the endoflagella, located beneath the outer membrane. Three genera of Spirochetes cause disease in humans: ***Borrelia, Leptospira*** and ***Treponema.***

spirochete	features	diseases	diagnosis	transmission
Borrelia burgdorferi	large, loosely coiled motile	Lyme disease	visible using a light microscope	bite from infected deer tick
Borrelia recurrentis	large, loosely coiled motile	relapsing fever	visible using a light microscope	human body lice are the vector humans are the reservoir
Leptospira interrogans	thin, tightly coiled motile	leptospirosis aseptic meningitis	can be grown on culture media visible by dark field microscopy	contact with urine of infected animals (dogs)

Treponema pallidum	tightly coiled
	motile
		syphilis
		STD
		  Primary
		  Secondary
		  Latent
		  Tertiary
		  Congenital
			cannot be grown on lab media
			visible by dark field microscopy
			non-specific serological tests:
			  VDRL, RPR
			specific-serologic tests:
			  FTA-ABS, MHA-TP
				sexually
				congenitally

Think **BLT** for bacon, lettuce and tomato; and all the Spirochete genera: ***Borrelia, Leptospira***, and ***Treponema.***

Borrelia

pathogen	lab findings	diseases	notes

Borrelia burgdorferi	large, **loosely coiled**
	
	motile
	
	Laboratory diagnosis looks for antibodies to the pathogen using ELISA.
	
	Visible by light microscope.
		Lyme disease (it was first described in Lyme, Connecticut)
		
		    Seasonal: occurs in summer
		
		    Transmitted by tick bites of the deer tick *Ixodes dammini* It takes the tick 24-48 hours to transmit the disease.
		
		    Early symptoms may include headache, stiff neck, fever, facial nerve palsies and a skin rash. Also present is a characteristic skin lesion called **erythema chronicum migrans** which looks like a target with a pale center and a red border.
		
		    Late complications include arthritis, neurological symptoms, meningitis and myocarditis.
		
		    Ticks are the vector and **small mammals** and **deer** are the reservoir.
			Lyme disease is the **most frequently reported tick borne illness** in the U.S.
			
			Antibiotic treatment depends on the symptoms and stage of the disease.
			
			Because it takes so long for the tick to transmit the disease, **inspecting the skin** for ticks after being outdoors may prevent infection (unfortunately the ticks are small and easily missed).

Borrelia recurrentis	large, **loosely coiled**
	
	motile
	
	Lab diagnosis involves seeing large spirochetes in a peripheral blood smear.
	
	Visible by light microscope.
		Relapsing fever
		
		    Transmission is person-to-person by the human body louse.
		
		    After entering the host, the pathogen causes fever, chills and lesions in the liver, GI tract, kidney and spleen.
		
		    The body produces antibodies, clears <u>most</u> of the organisms and within a week the fever is gone.
		
		    The remaining organisms then undergo **antigenic variation** of a protein known as **variable major protein** (VMP).
		
		    The **symptoms of infection** (fever, chills, headache) **relapse with each antigenic shift of VMP** which can occur 5-10 times.
		
		    Human body **lice** are the vector and **humans** are the only reservoir.
			B. recurrentis is the **only** *Borrelia* **species not transmitted by a tick.**
			
			Humans are the only host.
			
			Treat with doxycycline.

Think: **_Borrelia_ = _Big_**, only spirochete visible by light microscopy.

Gram Nothing

Leptospira

pathogen	lab findings	diseases	notes
Leptospira interrogans	thin, **tightly coiled**, long motile gram - like **obligate aerobes** Visible by **darkfield microscopy**. Only spirochete that can be grown on artificial media.	**Leptospirosis** **Transmission** occurs by contact with **urine** (swimming, drinking water) from infected animals. Organisms are carried in the blood and deposited throughout the body, particularly in the **kidneys**, **liver** and **meninges** where they multiply and cause organ damage. The disease is considered to be biphasic: 1) **Acute phase** corresponds to the presence of *L. interrogans* in the blood. Symptoms include: high fever, chills, malaise, myalgia and severe headaches. Severe cases may result in jaundice. 2) **Immune phase** correlates to the increase in antibody titer. Symptoms include: low grade fever and meningeal signs. The disease is self-limited and recovery usually occurs within two to three weeks.	*L. interrogans* is primarily a pathogen of animals. **Dogs** are the most frequent source of leptospirosis in the U.S. In dogs, rats and other animals, infection of the kidneys may be asymptomatic while the organism is shed continuously in the urine. **Vaccine** is available for domestic animals. Treat with penicillin or doxycycline.

Treponema

pathogen	lab findings and notes	diseases
Treponema pallidum	tightly coiled spirochete motile **Never been cultured** on any artificial media; grows only in animals or on cell culture. **Darkfield microscopy** is used to visualize organisms found in primary and secondary skin lesions. Diagnosed by direct and indirect serologic tests. Treat with **benzathine penicillin** a long acting antibiotic. Treat 1°, 2° and latent syphilis with IM benzathine penicillin. Treat 3° syphilis with IV penicillin.	**Syphilis** comes in two forms determined by the route of transmission: sexual or congenital. **Sexually transmitted disease** 50% of contacts with an infected partner result in transmission of the disease. **Primary syphilis** is characterized by a **painless**, hard chancre and swollen regional lymph nodes appearing **within 1-13 weeks** (mean of 3 weeks) of infection. The **chancre contains *T. pallidum*** and is infectious so contact transmission can occur. The chancre heals spontaneously. **Secondary syphilis** usually occurs 1-3 months after the chancre heals and is characterized by a maculopapular rash which appears on the skin (usually the palms and soles), genitalia and mucous membranes. The rash is often symmetric. The **rash contains the organism** and contact transmission can occur. Any organ in the body can be infected during secondary syphilis; commonly immune complex mediated **acute glomerular nephritis** and **arthritis** occur (Type III hypersensitivity reaction). The rash spontaneously resolves and the individual is no longer infectious. One quarter of all patients will have a recurrence of symptoms within two years. **Latent syphilis** Almost all cases of secondary syphilis will spontaneously resolve within 1-3 months of the initial infection and enter the latent phase. From the latent phase, the disease may take several different directions: 1. relapse back into secondary syphilis 2. progress to tertiary syphilis 3. remain asymptomatic Patients with latent syphilis will have no symptoms, but serology will be positive. Latent syphilis should be treated is if active (i.e. give the patient a penicillin shot).

Gram Nothing

pathogen	lab findings / notes	diseases
Treponema pallidum continued	**Jarisch-Herxheimer reaction**: Patients with secondary or tertiary syphilis may clinically degenerate (fever, chills, myalgia, tachycardia and flushing) within hours of **treatment**. This is due to the sudden lysis of the spirochetes which releases a bolus of antigens into circulation. This can be ameliorated with steroids.	**Tertiary syphilis** occurs in one third of untreated infections and may not present until 1-40 years after the initial infection. Lesions present all over the body:

Tertiary syphilis occurs in one third of untreated infections and may not present until 1-40 years after the initial infection. Lesions present all over the body:

Skin: **gummas** (granulomas) are painless swellings which eventually rupture to form an ulcer. Unlike primary and secondary lesions, tertiary lesions **contain very few organisms**.

Cardiovascular: The large blood vessels are damaged because the vasovasorum (small blood vessels which supply the cells of the larger blood vessels) are infected and obliterated. This can cause aortitis, **aortic regurgitation** and **aortic aneurysms**.

CNS (neurosyphilis) is a result of the spirochete invading the CNS; organisms can be found in the CSF.

> **Tabes dorsalis** is a neurologic deficit caused by the destruction of the dorsal roots of the spinal cord. It predominantly attacks the lower limbs. Symptoms include sharp pains, sensory neuropathy and loss of vibratory and position sense (proprioception). The autonomic fibers are also destroyed resulting in postural hypotension (sudden drop in blood pressure upon standing).

> **General paresis** is a result of invasion and destruction of brain parenchyma. Patients present with progressive **dementia**, **mania** and **tremors** of the face and tongue.

At one time, tertiary syphilis was a major cause of mental illness.

People with neurosyphilis frequently have **A**rgyll **R**obertson **P**upil.

> **A**ccommodation **R**eflex **P**resent — eyes move medially and pupils constrict when focusing on a close object

> **P**upillary **R**eflex **A**bsent — pupils fail to constrict in response to a bright light

Think: prostitutes get syphilis and on the job they **accommodate (ARP)** but don't **react (PRA)**.

Congenital syphilis occurs when the pathogen passes through the placenta into the fetus. Often causes fetal death.

> If the baby is born, it gets a **rash** followed by **sloughing of the skin** on hands and feet.

> Bone involvement often affects the **nose.**

> **Liver** damage may cause **jaundice.**

An infant can experience tertiary disease years later with blindness, deafness and neurosyphilis.

Laboratory diagnosis of syphilis

Two strategies are employed to determine the presence of syphilis:
- ◆ microscopic visualization of the spirochete from skin lesions
- ◆ serologic evidence of the body's reaction to syphilis (blood test)

Microscopic tests:

Examination of fluid from **skin lesions** of primary (chancre) or secondary (maculopapular rash) syphilis:

microscopic test	how the test works	comments
darkfield microscopy	Direct visualization of the organism. Darkfield microscopy is required to detect the very narrow spirochetes.	Not readily available. Requires skill and experience to interpret.
direct fluorescent antibody *T. pallidum* test (DFA-TP)	The test consists of mixing anti-*T. pallidum* antibodies conjugated to fluorescein with a sample. If the sample contains *T. pallidum*, the antibodies will bind and allow the Treponeme to be visualized with an ultraviolet microscope.	Alternative to darkfield microscopy. Commonly used.

Serologic tests.

Two types of antibodies are produced in *T. pallidum* infection:

① Nontreponemal (also known as reagin) antibodies are antibodies directed against a lipid antigen.
② Treponemal antibodies are directed against *T. pallidum* itself.

Thus, there are two classes of serologic tests which detect the presence of infection with *T. pallidum*:

① Nontreponemal: **RPR** (rapid plasm reagin), **VDRL** (venereal disease research laboratory).
② Treponemal: **FTA-ABS** (fluorescent treponemal antibody-absorption), **MHA-TP** (microhemagglutination assay for antibodies to *T. pallidum*).

Gram Nothing

Serological tests for diagnosing syphilis use either treponemal or non-treponemal antigens:

non-treponemal	how the test works	comments	% true positive
VDRL venereal disease research laboratory **RPR** rapid plasma reagin	The antigen for the VDRL and RPR tests is a mixture of cardiolipin, cholesterol and lecithin. Cardiolipin is obtained by grinding up cow hearts. In the presence of anti-treponemal antibody, cardiolipin flocculates (clumps up).	**Used as a screening test and to monitor the success of treatment.** Titers for VDRL and RPR rise following infection and parallel disease activity. As the disease improves, RPR and VDRL levels decrease and titers can be followed to monitor treatment. Titers also decrease in later stages of disease (latent and tertiary) so the test is not as reliable for these stages. **False positive results occurs in certain conditions.** Non-treponemal tests are nonspecific so conditions other than syphilis can induce non-treponemal antibodies and give a false positive result. Conditions associated with a false positive RPR and VDRL include viral hepatitis, infectious mononucleosis, malaria, IV drug addiction (narcotics), age greater than 70 years and autoimmune diseases such as systemic lupus erythematosus (SLE).	The sensitivity of this test falls over time. • primary 60-85 • secondary 100 • latent 75-90 • tertiary 45-90

treponemal	how the test works	comments
FTA-ABS fluorescent treponemal antibody-absorption	• The patient's serum is filtered to separate out any Treponemal antibodies which are not specific to *T. pallidum* (antibodies to other species of Treponemes other than *T. pallidum*). This reduces false positive results from Treponemes other than *T. pallidum*. • The patients serum is poured on to a slide covered with *T. pallidum* antigens. The slide is washed leaving the anti-treponemal antibody-antigen complexes on the slide. • Fluorescent antibodies which react with human immunoglobulins are added. These antibodies bind to the antibody-antigen complexes on the slide. • The slide is then examined with an ultraviolet microscope, any fluorescence indicates a positive test.	These antibodies <u>remain for life</u>. Once the test is positive it will always be positive. Expensive test. **% true positives** • primary 60-85 • secondary 100 • latent 100 • tertiary 100
MHA-TP microhemagglu-tination assay for antibodies to *T. pallidum*	• The patient's serum is filtered to separate out any Treponemal antibodies which are not specific to *T. pallidum* (antibodies to other species of Treponemes other than *T. pallidum*). This reduces false positive results from Treponemes other than T. *pallidum*. • *T. pallidum* (the antigen) is attached to sheep RBCs and is mixed with the patient's serum. If the serum contains antibodies to *T. pallidum*, the antibodies will bind to the *T. pallidum* antigens on the RBCs and cause agglutination. The detection of agglutination indicates a positive test.	These antibodies <u>remain for life</u>. Once the test is positive it will always be positive. **% true positives** • primary 65-85 • secondary 100 • latent 100 • tertiary 100

Notes

Gram Nothing

Gram Nothing

Chlamydia

Chlamydia are **obligate intracellular parasites** with a gram negative-like cell wall. They have a complex system for reproduction:

1. a small extracellular **elementary body** is phagocytosed by a host cell
2. the elementary body then differentiates into a **reticulate body** (also called initial body)
3. the reticulate body then reproduces by binary fission to form many new elementary bodies
4. the elementary bodies are released from the cell

Chlamydia are **anaerobic** and unable to produce ATP for their own energy needs so they are dependent on the host cell for energy. Their cell membranes contain ADP/ATP translocators which bring ATP into the bacteria and push ADP out of the bacteria. Since *Chlamydia* are intracellular parasites it is impossible to grow them on agar. They can only be grown in living cell cultures.

In infected tissues, *chlamydia* form **cytoplasmic inclusions** which are visible with iodine or Giemsa stains. The morphology of the inclusion bodies is used to differentiate the three species of chlamydiae.

species	inclusion morphology	sulfonamide sensitivity
pneumoniae	lacks glycogen Giemsa stain	?
psittaci	diffuse lacks glycogen Giemsa stain	-
trachomatis	compact contains **glycogen** iodine stain	+

Gram Nothing

The Microbiology Companion, Topf and Faubel ©1997

Chlamydia

obligate intracellular parasite

pathogen	diseases	notes
Chlamydia pneumoniae	**Atypical pneumonia** **Transmission** is person to person. Commonly causes pneumonia among college students.	There is no bird host involved. Treat with doxycycline or erythromycin.
Chlamydia psittaci	**Psittacosis** **Transmission** is by inhalation of organisms from the dry feces of infected **birds**. The organism can cause a latent infection in birds that reactivates if the bird has a stressful day. (See Tweety Bird for example of stressful bird day.) Infection occurs in the lung and can be asymptomatic or cause a serious, even fatal **pneumonia.** In severe infections, the disease spreads from the lungs to the liver and thyroid gland causing jaundice and thyrotoxicosis (excess thyroid hormone throughout the body). Infection of the meninges may result in delirium and death.	Treat with doxycycline or erythromycin. There is a high risk of infection for workers in poultry slaughter houses and exotic bird collectors.

Chlamydia trachomatis

obligate intracellular parasite

causes a wide variety of diseases

serotypes A-L

eye infection	disease characteristics	types, treatment and notes
Trachoma	*Chlamydia trachoma* causes two types of eye infections: trachoma and inclusion conjunctivitis. **Trachoma** is a chronic inflammation of the conjunctiva. Repeated infections cause scarring and may lead to **blindness**. When the scars contract, the eyelid folds under itself which causes the eye lashes to rub against the cornea. This mechanical abrasion predisposes to secondary bacterial infection which is the actual cause of blindness. **Transmission** is by fingers or fomites (i.e. towels) to the eye. **Relapses** occur due to the intracellular nature of the disease (i.e. it can hide out in a cell during treatment only to jump back out afterwards).	Caused by serotypes A, B, Ba, and C. Trachoma is treated with topical sulfonamides and tetracycline. Trachoma is widespread throughout the world and causes blindness in millions of people. It is especially common in Egypt and the Middle East.
Inclusion conjunctivitis	**Inclusion conjunctivitis** is an acute infection on the conjunctiva of newborns. **Transmission** occurs during passage through the birth canal of a mother with genital chlamydia. 5-12% of all pregnant women have a genital chlamydia infection and half of them will pass this on to the neonate in the form of inclusion conjunctivitis. Occurs one to two weeks after birth. Does not result in blindness.	Caused by serotypes D-K (which cause the genital infections). Inclusion conjunctivitis is treated with topical sulfonamides and tetracycline. It is <u>resistant</u> to the silver nitrate used for prophylaxis in newborns' eyes

Gram Nothing

Chlamydiae *trachomatis*

disease	disease characteristics	types, treatment and notes
Infant pneumonitis	**Infant pneumonitis** is an interstitial (as opposed to lobar) pneumonia of newborns.	Caused by serotypes D-K (which cause the genital infections).
	Transmission occurs during passage through the birth canal of a mother with genital chlamydia. 5-12% of all pregnant women have a genital chlamydia infection and 5-10% of them will pass this on to the neonate as infant pneumonitis.	Infant pneumonitis is treated with erythromycin of sulfonamides.
	Occurs 6-26 weeks after birth.	
	The pneumonia presents gradually with a persistent cough, shortness of breath and bilateral chest infiltrates on x-ray. The disease is mild and complete recovery is the rule.	

STD	disease characteristics	types, treatment and notes
Chlamydia	**Chlamydia** is the **most common sexually transmitted disease**.	Caused by serotypes D-K.
	In women, infection usually causes **cervicitis**. The infection may progress to salpingitis (infection of the fallopian tubes) or pelvic inflammatory disease.	Genital tract infections are treated with doxycycline , azithromycin or erythromycin. A one dose oral treatment with azithromycin is available and effective; use of this treatment assures compliance.
	In men, *C. trachoma* is the most common cause of **nongonococcal urethritis.**	
	Symptoms in both genders may be mild or absent.	
Lymphogranuloma venereum	**Lymphogranuloma venereum** is a common STD in South America.	Caused by serotypes L-1, L-2 and L-3.
	A papule forms around the genital area one to two weeks after infection.	Lymphogranuloma venereum is treated with doxycycline or erythromycin.
	The infection spreads to the local lymph nodes and the original lesion heals.	
	One or two months later, the nodes enlarge and become tender. The enlarged nodes are known as **buboes**, which may rupture and drain.	

Notes

Gram Nothing

The Microbiology Companion, Topf and Faubel ©1997 **72**

Rickettsiae

Rickettsiae are **obligate intracellular parasites** that can only replicate within other cells; growth on agar is not possible. Rickettsial diseases are characteristically diseases of animals which are transmitted to humans by an arthropod vector. The diseases are characterized by symptoms of fever, headache and a rash. Of notable exception is Coxiella burnetii which causes Q fever and has no arthropod vector and does not cause a rash.

The cell wall of Rickettsiae is similar to gram negative bacteria. Rickettsiae can be diagnosed by the Weil-Felix reaction. Rickettsiae produce soluble antigens which induce antibodies that cross react with certain strains of the bacteria Proteus. The patient's serum is mixed with three different Proteus strains (OX-19, OX-2, OX-K) and if the bacteria agglutinate the reaction is positive.

◆ All Rickettsiae species infect the **endothelium of blood vessels** which may cause **vasculitis**.
◆ All Rickettsiae species are susceptible to **doxycycline**.
◆ All Rickettsiae species are **transmitted by insect bites** except Q fever, *coxiella burnetii*.
◆ All Rickettsiae diseases **cause a rash** except Q fever, *coxiella burnetii*.

Family: *Rickettsiae*	vector	reservoir	diseases	OX-2	OX-19	OX-K	testicular swelling in guinea pigs
R. rickettsii	ticks	ticks, rodents and dogs	Rocky Mountain spotted fever	**+**	**+**	**-**	
R. akari	mouse mites	mice	rickettsialpox	**-**	**-**	**-**	
R. prowazekii	human body lice	humans / flying squirrels (rare)	epidemic louse-borne typus fever Brill-Zinsser disease	**-**	**+**	**-**	**-**
R. typhi	rat fleas	fleas and rats	endemic flea-borne typhus	**-**	**+**	**-**	**+**
R. tsutsugamushi	rat mites	mites and rats	scrub typhus	**-**	**-**	**+**	
Coxiella burnetii	**spores** (no vector)	cattle, sheep, goats, cats, humans	Q fever	**-**	**-**	**-**	

pathogen	appearance and lab findings	diseases	notes/important properties
Rickettsia rickettsii	gram negative like Weil-Felix reaction: OX-19 positive OX-2 positive OX-K negative	**Rocky Mountain spotted fever** (RMSF) is most commonly found in the southeast. Transmission is by hard shelled, slow feeding *Dermacentor andersoni* ticks (wood tick) or *Dermacentor variabilis* ticks (dog tick); it takes these ticks 4 hours to transmit the disease. RMSF is characterized by fever, headache, and a maculopapular **rash** (the spotted part of the name) which **starts on the hands and feet** and spreads to the trunk. Infection occurs most often in children and must be differentiated from meningococcemia. The organism grows in the endothelium of small blood vessels and causes **vasculitis**. **Ticks** are the vector and **rodents, tick** and **dogs** are the reservoir.	Treat with doxycycline or chloramphenicol. **Transovarian** passage of the organism occurs in the tick.
Rickettsia akari	Weil Felix reaction: negative	**Rickettsialpox** is transmitted by the bite of an infected mite. Initially, a lesion occurs at the site of the bite. About two weeks later, fever, chills and a **rash** similar to chicken pox occur. Recovery is spontaneous. **Mouse mites** are the vector and **mice** are the reservoir.	Treat with doxycycline or chloramphenicol.

Gram Nothing

pathogen	appearance and lab findings	diseases	notes
Rickettsia prowazekii	Weil Felix reaction: OX-2 negative **OX-19 positive** OX-K negative	**Epidemic louse-borne typhus fever** is associated with poor sanitation and ravages the populace during **wartime**. The organism multiplies in the GI tract of the **human body louse** which transmits the disease in its **feces**. The louse defecates while biting someone. The bite is irritating and the host scratches the feces into the bite. The organism grows in the endothelium of blood vessels and causes **vasculitis**. The disease is characterized by a high fever (104°F) and a **rash** which **starts on the trunk** and moves to the extremities. Neurological symptoms such as stupor or delirium may be present. The patient either recovers or dies within 15-20 days. **Brill-Zinsser disease** is a sporadic and mild form of typhus fever which only occurs in people who have recovered from epidemic typhus. Latent organisms are responsible for this disease which is mild due to the amnestic response of the immune system. Latent infection is a reservoir of disease between epidemics. Human body **lice** are the vector and **humans** are the reservoir.	**No transovarian** passage of the organism occurs in the louse. There is a rare sporadic form that is transmitted to humans by fleas (vector) from a **flying squirrel** (Rocky) reservoir. Epidemics are best controlled by world peace, delousing and good sanitation. Treat with doxycycline or chloramphenicol. *R. typhi* and *R. prowazekii* are so closely related that getting either disease gives resistance to the other.

pathogen	appearance and lab findings	diseases	notes/important properties
Rickettsia typhi	Weil Felix reaction: OX-2　negative **OX-19　positive** O-K　negative Causes **testicular swelling** if inoculated intraperitoneally in a male guinea pig. *R. prowazekii* causes no swelling.	**Endemic flea-borne typhus or murine typhus** Clinically indistinguishable from the louse borne epidemic form caused by *R. prowazekii* but sequelae are less severe. The **flea** is the vector and the **rat** is the reservoir. Occurs most commonly in Texas and California.	*R. typhi* and *R. prowazekii* are so closely related that infection by one induces immunity to the other. Treat with doxycycline or chloramphenicol. **Transovarian** passage of the disease occurs in the flea.
Rickettsia tsutsugamushi	Weil Felix reaction: OX-2　negative OX-19　negative **OX-K　positive** (some varieties are negative)	**Scrub typhus** is endemic in Asia and the South Pacific and is transmitted by the bite of a **mite**. The incubation period is 1 to 2 weeks. Symptoms of infection include headache, fever and a lesion at the site of the bite. Can cause latent infections which can reactivate in the future. **Rat mites** are the vector and **rats** are the reservoir.	Treat with doxycycline or chloramphenicol. **Transovarian** passage of the disease occurs in the mite.

Gram Nothing

Rickettsiae

The Microbiology Companion, Topf and Faubel ©1997 **76**

pathogen	appearance and lab findings	diseases	notes/important properties
Coxiella burnetii	Weil Felix reaction: negative Diagnosis is by the explosive on-set of symptoms and a 4X increase in titers.	**Q fever** is a respiratory tract infection. **Transmission** is by **inhalation** of dust contaminated with the byproducts of infected animals. Symptoms of infection include fever, headache, and malaise (no rash). The disease often progresses to **atypical pneumonia**. May produce a latent infection which can be reactivated by immune suppression. These late infections often cause culture negative endocarditis or hepatitis. The organism infects many **animals (sheep, cattle**, etc.) which serve as **reservoirs** for infection. **There is no arthropod vector.**	*Coxiella burnetii* is resistant to drying and heat and is therefore stable enough extracellularly to be transmitted through the environment. Treat with doxycycline or erythromycin. Think: Q stands for queer: • no arthropod vector • disease does not cause a rash • organism can survive in environment

Notes

Gram Nothing

Mycoplasma

These bacteria are the smallest free-living organisms. They do not have a cell wall.

pathogen	appearance and lab findings	diseases	molecules and enzymes	notes
Mycoplasma pneumoniae	smallest free-living organisms **no cell wall** stains poorly with gram staining **Beta hemolytic**: clear zone around colonies. Only bacteria which contain **cholesterol** in their membrane and require cholesterol for growth. The colonies formed on artificial media are so small that a micro-scope is required to see them. Colonies look like **fried eggs**. Diagnosis is by the appearance of **cold agglutinins** which appear in sera. Cold agglutinins are IgM antibodies which agglutinate RBCs at low temperatures but not at body temperature (37°).	**Atypical pneumonia** (walking pneumonia) commonly occurs in school children and persons in military camps. **Transmission** is via respiratory droplets and the incubation period is 2-3 weeks. Prolonged shedding of the organism occurs for up to 14 weeks. Infection is characterized by fever, chills, headaches and a nonproductive cough. Neurological symptoms are common and help to differentiate *Mycoplasma* from *Chlamydia* or *Legionella* (other causes of atypical pneumonia). Severe infections usually occur in older people. Humoral immunity is not protective. Repeat infections are possible and result in a more severe disease which may be autoimmune mediated.	Contains a receptor which allows it to bind to RBCs and tracheal epithelial cells. Produce **hydrogen peroxide** as a byproduct of metabolism. This is the agent which causes hemolysis of RBCs.	Causes disease only in humans. Treat with erythro-mycin or azithro-mycin. Mycoplasma do not have a cell wall so cell wall antibiotics (e.g. penicillin, vancomycin) are ineffective.

Think: **old** and **cold**.   **Old**: came out before the development of cell walls.

Cold: cold agglutinins appear in sera during infection.

IgM: the middle of the M is a down arrow indicating low temperature, cold agglutinins.

Actinomycetes

Actinomycetes are true bacteria (the suffix mycetes is *usually* reserved for fungi) that demonstrate **branching filamentous growth** which **resembles** fungi. **Actinomycetes are not fungi**. Two genera of Actinomycetes cause disease in humans: *Actinomyces* (gram +) and *Nocardia* (acid fast rod).

pathogen	appearance and lab findings	diseases	notes/important properties
Actinomyces israelii	gram + obligate anaerobe	**Actinomycosis** infection occurs endogenously in areas of trauma, usually in the head, neck, abdomen and chest. Infection is characterized by the formation of **sulfur granules** which are yellow masses of filaments contained in abscesses draining through **sinus tracts** in the skin. Untreated infection can cause severe organ and/or **bone damage**. The disease is not communicable. Dental carries may be contributed to by *Actinomyces*.	**Normal flora** of the oral cavity and GI tract. *A. israelii* infections are associated with the use of IUDs. Treat with penicillin G or ampicillin for at least a year. Surgical excision of the abscess may be necessary.
Nocardia asteroides	acid fast rod aerobes	**Nocardiosis** is a **pulmonary infection** which primarily occurs in **immunocompromised** persons. **Transmission** is by inhalation of the organism. Infection is characterized by the formation of **lung abscesses** and sinus tracts without sulfur granules. Lung abscesses form cavities similar to TB. When an abscess erodes into a blood vessel, the organism can spread via the blood to any organ, particularly the brain and kidney. Bone damage is rare. The disease is not communicable.	*Nocardia asteroides* is found in **soil**. Treat with trimethoprim/sulfamethoxazole. Prognosis is poor if the organism spreads throughout the body.

The Microbiology Companion, Topf and Faubel ©1997

Gram Nothing

Mycobacteria

Mycobacteria are acid fast bacteria which cause 2 major diseases in humans: tuberculosis and leprosy.

Mycobacteria are not considered to be gram positive or gram negative because the bacteria do not stain with gram staining. Instead they are considered to be **acid fast**. Mycobacteria are stained using the Ziehl-Neelsen technique in which a slide with the bacteria is covered with the red stain, carbolfuchsin and then heated for several minutes to allow penetration of the dye into the cell wall. The smear is then washed with a solution of 95% alcohol and 3% hydrochloric acid and counterstained with methylene blue. Acid fast organisms are not decolorized by the acid alcohol solution (hence the term acid fast) and will appear red; all other organisms will appear blue. The high lipid content of the Mycobacteria cell wall (about 60%) confers its acid fast property.

Mycocides are a particular type of lipid which are unique to acid fast organisms. Mycosides consist of a type of mycolic acid (which differs among the Mycobacteria species) attached to a carbohydrate. **Cord factor**, a type of mycocide, is the only known virulence factor for Mycobacteria and causes the bacterium to grow in parallel cords. It also inhibits the migration of PMNs, disrupts cellular respiration and induces the production of tumor necrosis factor (TNF). TNF is a lymphokine (something released by lymphocytes) which is responsible for the cachexia seen in patients with cancer and TB.

pathogen	lab findings	diseases	important properties
Mycobacterium tuberculosis also known as the tubercle bacillus	acid fast bacilli (AFB) **obligate aerobe** produces **niacin** **Cord factor** is a mycocide and is associated with virulence. It induces the organism to grow in parallel cords. **Wax D** (a mycocide) and **muramyl dipeptide** are adjuvants. (An adjuvant causes an enhanced immune response when injected with antigen. Adjuvants are used in vaccines to ↑ effectiveness) These adjuvants are thought to be involved in the development of delayed type hypersensitivity (type IV) and granuloma formation. **Intracellular** growth within reticuloendothelial cells.	**Tuberculosis** (TB) is transmitted by **respiratory droplets** from someone with active pulmonary disease. Rarely infection occurs through skin contact or ingestion of organisms. **It is highly contagious.** **Primary tuberculosis** occurs after the first encounter with *M. tuberculosis.* After inhalation, the organisms are phagocytized by macrophages where they reproduce intracellularly. The infected macrophages drain into the hilar lymph nodes and other parts of the body. 2-6 weeks later, cell mediated immunity responds by walling off the infected macrophages into a **tubercle**. A tubercle is a caseous **granuloma** which contains Langhans cells (multinucleated giant cells), epithelioid cells and necrotic tissue surrounded by a ring of lymphocytes. The tubercle and draining lymph node together are known as a **Ghon complex**. The Ghon complex undergoes progressive fibrosis and heals by calcification which is visible on X-ray. *M. tuberculosis* organisms within the tubercle may remain viable for decades. Usually occurs in the **lower lobes**. The infection is usually asymptomatic. Persons with primary tuberculosis are infected, but do not have active disease. They are therefore not a source of infection and cannot transmit the organism to others. Occasionally, the primary lesion does not heal and primary TB immediately goes on to secondary TB	**Treatment** with multi-drug therapy is necessary to prevent the development of drug resistant mutants. **Prognosis** is usually good when disease is localized to the lungs. Prognosis becomes dismal, however, in patients infected with drug resistant strains. **Isoniazid** (INH) and **rifampin** are commonly used simultaneously with other drugs for 6-9 months. **Control of TB** relies heavily on **prevention**. Therapy with **INH** is recommended for individuals with the potential for developing active disease such as persons with a TB skin test that has recently turned positive or those have close contact with a person recently diagnosed with tuberculosis (family members). INH can cause liver damage in persons over 35. Therefore, age must be consideration in the decision to prescribe INH.

Gram Nothing

pathogen	lab findings	diseases	notes
Mycobacterium tuberculosis also known as the tubercle bacillus continued from previous page	Stained using the **Ziehl-Neelsen technique** which is a method for staining acid-fast organisms. Cultures are prepared on **Löwenstein-Jensen medium**. **Slow growth**: 6-8 weeks for visible colonies. Each cell division takes 12-20 hours. Diagnosis is by observing AFB in sputum samples. **Resistant to drying** and many chemical agents i.e. acids, bases. Sensitive to heat, pasteurization kills tubercle bacilli.	**Secondary tuberculosis (reactivation tuberculosis)** occurs when organisms latent within a healed primary lesion (tubercle) reactivate and cause infection within the lung. Reactivation occurs as a result of depressed immunity and is common in people with AIDS, homeless persons, and the elderly, or in response to stress. Reactivation and spread usually occurs in areas of the body with a **high O$_2$** content: the **upper lobes of the lung**, **kidneys**, and **brain**. The secondary lesions (reactivation sites) of the lung enlarge and erode through the bronchial walls. Tubercle bacilli is contained in the sputum which becomes a source of infection. The majority of cases of active TB are thought to represent reactivation of a healed primary infection. **Early symptoms** include fever, malaise, loss of appetite and weight loss. **Later**, **cough**, hemoptysis and pleuritic pain may develop. **Tuberculosis meningitis** is a lethal complication which results from invasion into the CNS. **Miliary tuberculosis** is a disseminated infection which may occur after primary infection or reactivation. It is characterized by small nodular lesions throughout the lung or the whole body. **Pulmonary**: the organism drains through lymphatics into the pulmonary circulation and is seeded throughout the lung. Small lesions can be seen scattered in the lung tissue on X-ray. **Systemic**: lesions erode through blood vessel walls allowing hematogenous (in the blood) spread via the systemic circulation. The organism disseminates throughout the body; almost any organ may be involved.	Only a small percent of those infected with *M. tuberculosis* will actually develop active disease. **BCG vaccine** is used in other countries but not the U.S. It is a **live vaccine** with questionable effectiveness. It converts the recipient to a positive skin test. **Tuberculin skin test** measures delayed hypersensitivity (type IV) to purified protein derivative (PPD). A positive reaction is 10 mm of induration and erythema. The test is not positive until 2-6 weeks after infection. A positive test only indicates that the person is infected, not whether the disease is active or inactive. False negatives (anergy) occur in patients with severe disseminated disease or AIDS. Worldwide, there are 8-10 million new cases and 3 million deaths every year. It is estimated that **1 billion people are infected.**

Primary Tuberculosis

Infection begins after the patient inhales tu-berculosis bacilli into the lungs, most com-monly the lower lobes. The normal inflammatory response does not destroy the bacilli.

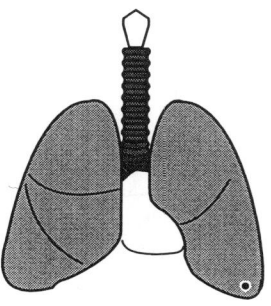

Activated T-cells circulate through the body. This marks the acquisition of delayed type hypersensitivity (type IV) and patients will have a positive PPD skin test. When activated T-cells come in contact with bacilli in the lung, they release chemotactic factors (gamma interferon) which recruit monocytes to form a granuloma

Some of the bacilli are engulphed by alveolar macrophages and drain to local lymph nodes (hilar and tracheobronchial). In the nodes, bacilli antigens are presented to antigen specific helper T-cells. These cells release lymphokines which cause clonal expansion of T-cells.

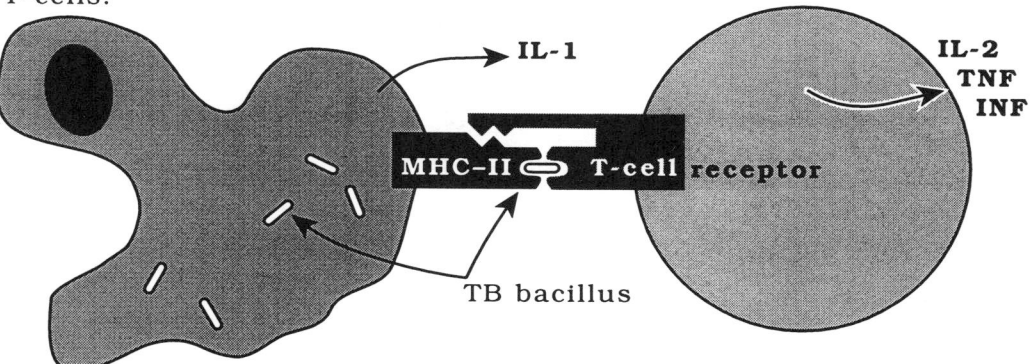

The body walls off the tuberculosis bacilli is a unique scar called a granuloma. Their are multiple elements which make up the granuloma:

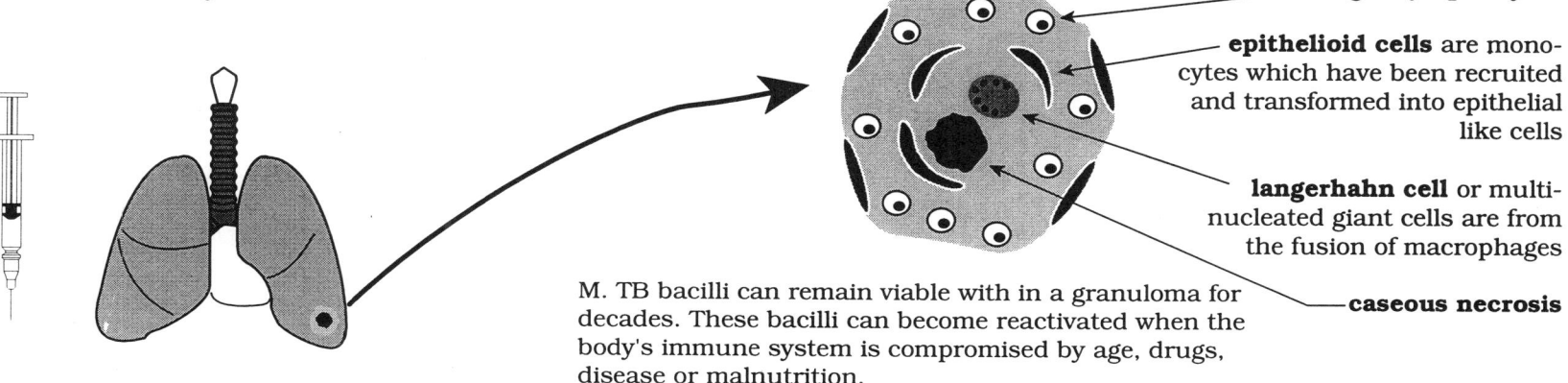

ring of **lymphocytes**

epithelioid cells are monocytes which have been recruited and transformed into epithelial like cells

langerhahn cell or multinucleated giant cells are from the fusion of macrophages

caseous necrosis

M. TB bacilli can remain viable with in a granuloma for decades. These bacilli can become reactivated when the body's immune system is compromised by age, drugs, disease or malnutrition.

Mycobacteria tuberculosis

Gram Nothing

Secondary Tuberculosis

Secondary tberculosis occurs in less than 5% of the people who have primary TB. It usually occurs years after the initial infection but it can occur immediately following the primary inoculation.

Secondary infection usually occurs in the apex of the lung due to the higher oxygen content there.

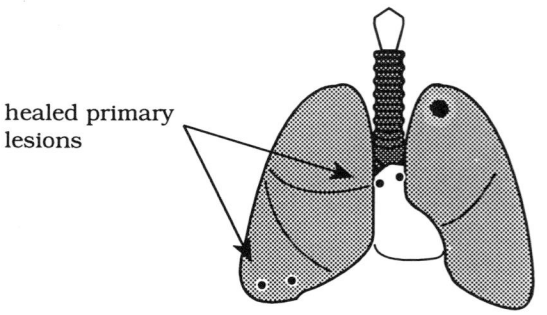

The granulomas can erode through the walls of the bronchi allowing the patient to cough up infectious bacilli and spread the disease.

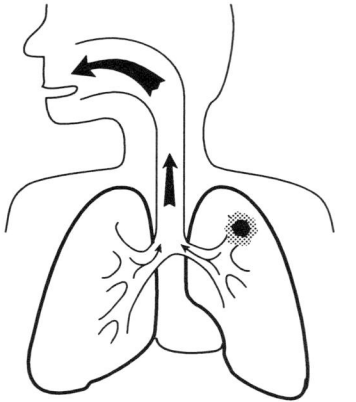

Disseminated Tuberculosis

Pulmonary Milliary Tuberculosis occurs when the lesion cavitates through a lymphatic channel. The organisms travel through the thoracic duct and into the right atrium. From the right heart they enter the pulmonary artery and go back into the lung. These lesions can grow and destroy entire lobes of the lung.

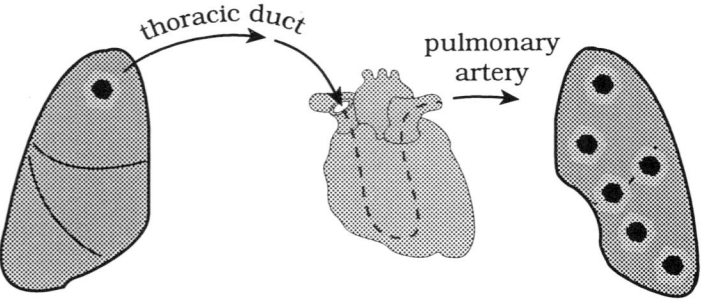

Systemic Milliary Tuberculosis occurs when the lesion cavitates into the pulmonary veins. The organism is then bought back to the left side of the heart and pumped out through the aorta to the entire body. This results in the seeding of the bacilli in any organ with a high oxygen tension: meninges, bone marrow, kidneys and adrenals.

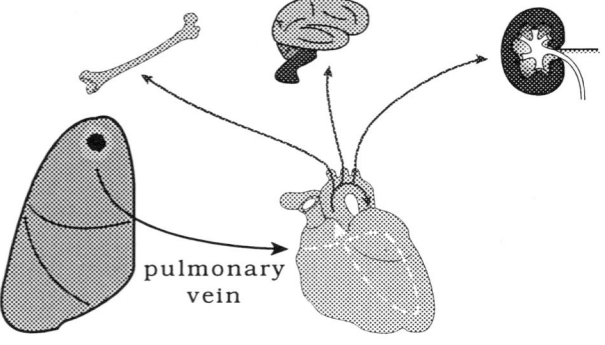

Secondary Tuberculosis

Secondary tuberculosis occurs in less than 5% of the people who have had primary TB. It usually arises years after the initial infection when the host's defenses are compromised but it can occur immediately following the initial infection. The secondary lesion is met with granulomatous inflammation immediately because of prior sensitization. While the source of infection is from the 1° lesion in the lower lobes the 2° is most often at the apex of the lung. The lesion usually cavitates into a bronchus allowing the bacilli to be expired and transmitted to others.

The secondary lesion can continue to grow and reduce the amount of functioning lung tissue. It is the body's response to the bacteria which causes the most damage.

The disease can disseminate in a few different ways. **Pulmonary Miliary disease** occurs when the lesion cavitates through a lymphatic channel. The organisms travel through the lymphatics and enter the right heart via the thoracic duct. From the heart, the bacilli go back to the lungs where they seed the entire organ. These lesions can grow and destroy entire lobes of the lung.

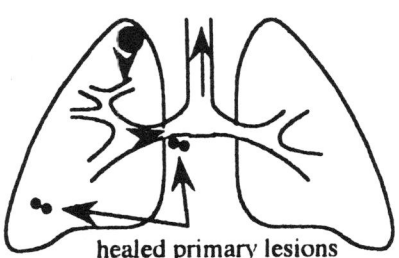

healed primary lesions

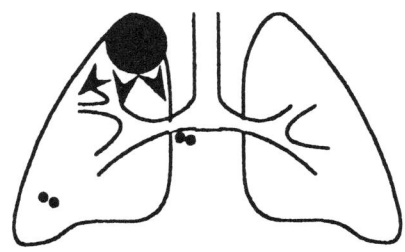

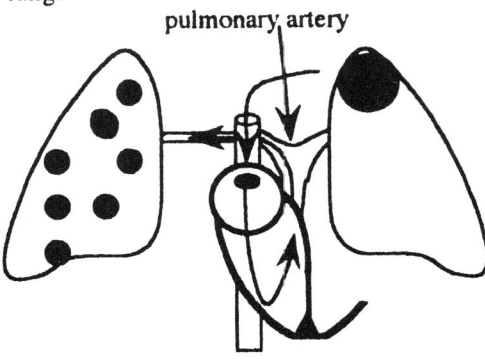

pulmonary artery

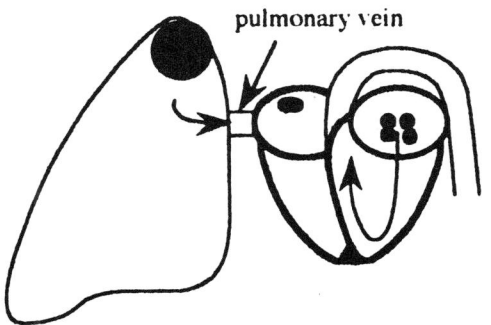

pulmonary vein

In **systemic miliary tuberculosis** the lesion cavitates into the pulmonary veins. The organism is then brought back to the left heart and pumped out in to the systemic circulation. This results in seeding of the bacilli in any organ with a high oxygen tension: meningis, bone marrow, kidneys and adrenals.

Gram Nothing

pathogen	appearance and lab findings	diseases	notes/important properties
Mycobacterium leprae	acid fast rods Has **never been grown** on artificial media or in cell culture. Can be grown on mouse footpad and can cause infection in armadillo. Serological diagnosis by ELISA is available. **Intracellular** growth within skin histiocytes and Schwann cells.	Infection causes two clinically distinct forms of leprosy: **lepromatous** and **tuberculoid**. Both forms of leprosy cause disfiguration secondary to skin anesthesia, bone resorption and thickening of the skin: **Lepromatous leprosy** is a progressive and **chronic** disorder which usually kills the patient. The body mounts a **poor cellular immune response** so that **many** organisms are found in every organ of the body. The disease attacks the **skin** primarily to cause **leonine facies** (lion-like); blindness and nose deformities are also common. The nerve damage is often bilaterally symmetric. **Negative** lepromin skin test. Possibly spread by respiratory droplets, the organism is found in nasal secretions. The **incubation period is very long** (3-10 years). **Tuberculoid leprosy** is often a **self-limiting** disease. This is due to the fact that it induces a **normal cellular immune response**. **Few organisms** are found in the granulomatous lesions. It primarily infects **nerves** which are damaged by the cellular immune response to the bacteria. This causes areas of **anesthesia** on the skin. **Positive** lepromin skin test. The incubation period is very long (3-6 years).	**Humans** are the only source of infection. The **optimal temperature** for growth is below normal body temperature, so the organism preferentially infects skin and superficial nerves. Children are more susceptible than adults and twice as many men are infected as women. Most people are naturally immune, but susceptible individuals require only brief contact with the disease to get infected. **Lepromin** is a purified protein derivative, analogous to tuberculin, which is used to test for the disease. It is of little value because persons with lepromatous leprosy give a negative reaction and persons who are PPD positive are also lepromin positive. Treatment is long term: from months to years.

Atypical Mycobacteria are unlike *Mycobacteria tuberculosis* and are normally found in soil and water. They cause a variety of pulmonary infection similar to, but usually milder than, *M. tuberculosis.* The atypical Mycobacteria can be divided into four categories based on their ability to produce pigments in varying conditions at various growth rates:

classification	growth rate	pigment production	example
Photochromogens	slow	yellow pigment only when grown in the light	*M. kansasii* *M. marinum*
Scotochromogens	slow	orange pigment when grown in the light or dark	*M. scrofulaceum*
Nonchromogens	slow	no pigment under any circumstances	*M. avium intracellularis complex*
Rapid growth	fast	no pigment under any circumstances	*M. fortuitum complex*

Gram Nothing

pathogen	appearance	diseases	important properties
Mycobacterium kansasii	**photochromogen** acid fast	**Pulmonary infection** is clinically similar to tuberculosis. Occurs most frequently in Texas.	**Treat** with rifampin, ethambutol, and INH. *M. kansasii* is antigenically similar to *M. tuberculosis* and cross reacts with PPD. Persons with *M. kansasii* infection will have a positive PPD skin test.
Mycobacterium marinum	**photochromogen** acid fast	**Draining, granulomatous ulcers** form at sites of trauma or abrasion which are infected by the organism. *Mycobacterium marinum* is found in fresh, salt and swimming pool **water**.	Surgical excision of granulomas. may be necessary. Antibiotic treatment includes rifampin and ethambutol.
Mycobacterium scrofulaceum	**scotochromogen** acid fast	**Scrofula** is a granulomatous inflammation of the lymph nodes which usually effects the cervical nodes. The organism is found is soil and water with a worldwide distribution.	Treat with surgical excision of affected lymph nodes.

Mycobacterium avium-intracellularis complex	**nonchromogen**
	
	acid fast
	
	consists of two species:
	  M. avium
	  M. intracellularis
		Pulmonary infection is clinically similar to tuberculosis.
		
		Infection occurs in **immunocompromised** persons, and is the **most common systemic bacterial infection** seen in persons with **AIDS.**
		
		Mycobacterium *avium-intracellularis* is found in water and soil especially in the Southeastern U.S.
		
		Mycobacterium avium-intracellularis infects birds and other animals.
			Because of resistance, several anti-tuberculosis drugs are used in combination to treat infection.

Mycobacterium fortuitum complex	**rapidly growing**
	
	acid fast
	
	includes:
	  M. fortuitum
	  M. chelonei
	  M. abscessus
		Wound infections present as skin abscesses or deeper infections.
		
		Infection may occur following surgery.
		
		The organisms are found in soil and water.	

Section two: Viruses

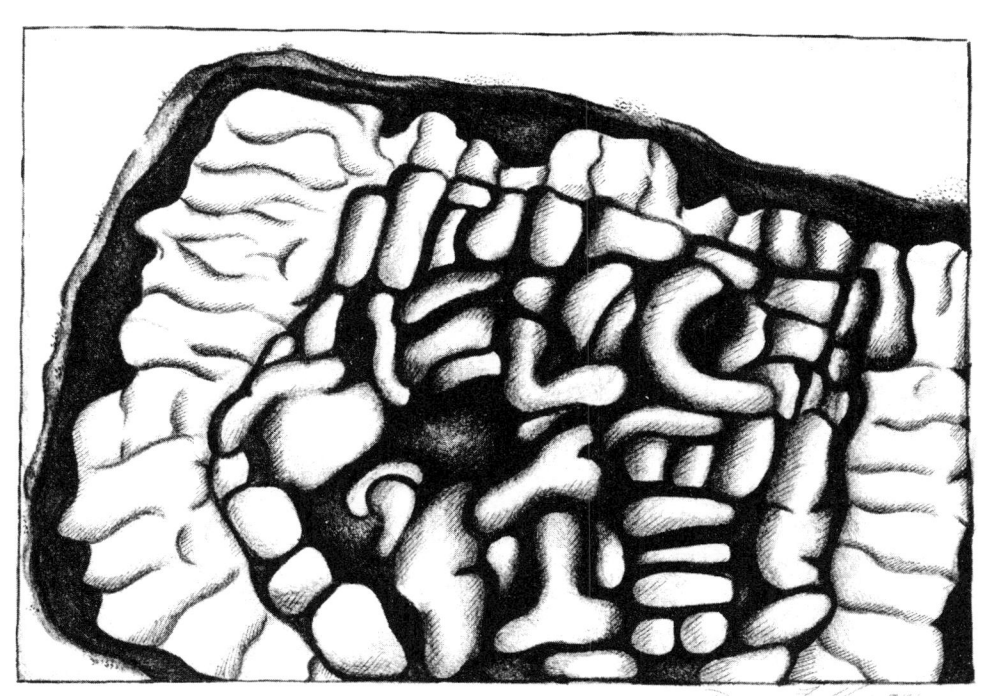

Small Pox, Dennis Franklin

Viruses

Viruses are the smallest form of life. They consist of small amounts of nucleic acid wrapped in a shell of protein and sometimes a lipid envelope. Viruses completely depend on host cells to replicate and are obligate intracellular parasites. Some major differences between viruses, bacteria and the other obligate intracellular parasites *Chlamydiae* and *Rickettsiae*, are shown below.

Characteristic	Bacteria	Chlamydia	Rickettsiae	Viruses
Nucleic acids	Both DNA and RNA.	Both DNA and RNA.	Both DNA and RNA.	**Either** DNA or RNA.
Walls and membranes	Cell wall made of peptidoglycan (except *Mycoplasma*). Cell membrane made of lipid bilayer with proteins imbedded in it.	Gram - like cell wall. Cell membrane made of lipid bilayer with proteins imbedded in it. Has ADP/ATP translocator to bring ATP into the bacteria and pump ADP out.	Gram - like cell wall. Cell membrane made of lipid bilayer with proteins imbedded in it.	Protein capsid which self-assembles around the viral genome. Can be icosahedral (20-sides), helical (cylindrical coil) or complex. Some viruses have lipid membrane or envelope.
Reproduction	Binary fission: one bacteria divides into two bacteria, each being smaller than the original.	Only able replicate in host cells. The reticulate body undergoes binary fission repeatedly to produce many elementary bodies (extracellular form) which are released.	Only able replicate in host cells.	Only able to replicate in host cells. Reproduce by assembly. The viral genome is translated by the host cell to produce many copies of the virus (both protein coat and nucleic acid genome). One virus can produce hundreds of progeny at once by mass production.
Enzymes	Bacteria have many enzymes which are necessary to produce ATP, reproduce, etc.			Very few enzymes, some viruses carry enzymes to initiate nucleic acid replication.

Viruses are classified by various attributes. The major division is whether the genome is RNA or DNA. Viruses are also identified by the presence or absence of a lipid envelope. Further classification is based on the characteristics of the genome: double or single stranded, circular or linear, segmented or single piece and whether circular DNA is supercoiled or not.

⊃ For a review of antiviral medications, please see *The Pharmacology Companion*, pages 191-196.

DNA Viruses

DNA Viruses

There are six families of DNA viruses which are categorized by the presence or absence of a lipid envelope, the type of DNA (single stranded or double stranded; circular or linear), and the shape of the capsid (complex or icosahedral). The majority of DNA viruses have double stranded DNA (except parvo) that is linear (except papova and hepadna) with an icosahedral capsid (except pox).

non-enveloped	enveloped
Papovavirus DS **circular supercoiled** icosahedral	**Hepadnavirus** DS **circular incomplete** icosahedral
Adenovirus DS linear icosahedral	**Poxvirus** DS linear **complex**
Parvovirus **SS** linear icosahedral	**Herpesvirus** DS linear icosahedral

non-enveloped	enveloped
Papovavirus Papilloma JC BK SV40 Polyomavirus	**Hepadnavirus** Hepatitis B virus (HBV)
Adenovirus Adenovirus	**Poxviruses** Smallpox virus Vaccinia virus Molluscum contagiosum virus
Parvovirus B19	**Herpesvirus** Herpes Simplex I Herpes Simplex II Cytomegalovirus (CMV) Epstein-Barr virus (EBV) Varicella-Zoster virus (VZV)

General rules about the life cycle of a DNA virus

Transcription of mRNA occurs in the host cell nucleus and uses <u>host cell DNA dependent RNA polymerases</u>, except <u>poxvirus</u> which carries its own DNA dependent RNA polymerase and never enters the nucleus.

    mRNA is processed before entering the cytoplasm for translation.

        Processing of viral mRNA includes:

            Splicing out introns.

            Capping the mRNA.

        Processing allows some viruses to produce multiple proteins from one strand of RNA through alternative splicing and shifting the reading frame.

    Two types of mRNAs:

Early

        Encode proteins required for DNA replication.

        Encode proteins necessary for expression of other genes.

        Cancer causing DNA viruses only express the early mRNA in transformed (neoplastic) cells. Late mRNA is never expressed.

Late

        Encode structural proteins such as capsid and membrane proteins.

        Since the viral DNA genome has already been replicated, late mRNAs are usually transcribed from this new DNA.

Translation into proteins occurs in the cytoplasm using host cell ribosomes.

Replication of DNA occurs in the nucleus, except for Poxvirus. Replication of DNA is by <u>viral DNA dependent DNA polymerase</u> (an early mRNA protein).

Assembly of the viral DNA with the viral capsid, may occur in the cytoplasm or nucleus.

    It is a spontaneous process not requiring enzymes or energy.

    The enveloped viruses get their membrane by <u>budding:</u>

        Through the cell membrane (Hepadna and Pox).

        Through the nuclear membrane (Herpes). This usually does not damage the cell.

    The nonenveloped viruses are released by:

        <u>Lysis</u> of the cell (killing the host cell.)

        <u>Extrusion</u> from the cell (does not destroy the cell.)

Viral Hepatitis

Hepatitis is a liver infection caused by many different unrelated viruses that will be discussed with their appropriate families. Despite the various etiologies the clinical picture is relatively consistent. Acute infection is characterized by the following signs and symptoms:

LUNCH	With	JAVA	And	Food
Lassitude	Weakness	Jaundice	Alanine and Aspartate Amino-	Fever
Urine (dark, due to ↑ bilirubin)		Anorexia	transferase (≈10x normal)	
Nausea		Vomiting		
Chills		Abdominal pain		
Headache				

virus	family	structure	incubation	transmission	complication
Hepatitis A virus	Picornaviridae Enterovirus	SS (+) RNA icosahedral capsid nonenveloped	2-6 weeks	fecal-oral	none, severity increases with age
Hepatitis B virus	Hepadnavirus	DS circular DNA icosahedral capsid envelope	7-26 weeks	blood borne disease; STDs, injection drug use, blood transfusion	long-term carrier state, cirrhosis, hepatocellular cancer
Hepatitis C virus formerly non A-non B	Flavivirus-like	SS (+) RNA icosahedral capsid envelope	2-4 weeks	blood borne disease; most common cause of post-transfusion hepatitis	chronic persistent infections, not as severe as HBV
Delta virus	defective virus; requires HBV for transmission	SS RNA genome with a HBV coat: icosahedral capsid envelope	super-infection 2-6 weeks co-infection 6-10 weeks	blood borne	none in co-infection; increased mortality in super-infection
Hepatitis E virus	Calicivirus	SS RNA nonenveloped	3-7 weeks	water borne epidemics	causes epidemics in Asia, no chronic infections reported

In addition to the above viruses, other viruses can occasionally cause hepatitis: Epstein-Barr virus, cytomegalovirus and herpes simplex virus.

Hepatitis B Virus (HBV)

of the Hepadnavirus family

characteristics	life cycle	serum antigens
enveloped icosahedral **DS, circular, incomplete DNA** The positive strand of the DNA is short and incomplete while the negative strand (complimentary strand) is long and complete. The positive strand fills in the gaps with a DNA polymerase using the (-) strand as a template. **Unable to be grown in cell cultures.** DNA dependent DNA polymerase carried within the core. RNA dependent DNA polymerase carried within the core (reverse transcriptase). Stable virus, can survive and remain infectious 6 months at room temperature.	1. Entry and uncoating; viral DNA travels to the **nucleus**. 2. In the nucleus, the DNA dependent DNA polymerase completes the synthesis of (+) DNA using the (-) strand DNA as a template. 3. (+) mRNA is transcribed from the completed DNA and then travels to the cytoplasm for translation into viral proteins. 4. The (+) mRNA is packaged into the nucleocapsid where **reverse transcription** occurs: (-) DNA is synthesized from (+) mRNA (this is RNA dependent DNA synthesis). The mRNA is destroyed at this time. 5. Incomplete (+) DNA is synthesized from (-) DNA strand. 6. Envelope is obtained by budding through the **cell** membrane. There is some evidence that viral DNA is inserted into hepatocyte DNA. This might explain the cancer and chronic disease associated with HBV infections.	**HBsAg**: Sera of infected individuals contains large quantities of **22 nm** particles that are either spherical or filamentous in shape. These particles are composed of **surface antigen**. **Dane particles**: Sera of infected individuals also contains smaller quantities of **42 nm** particles which are **intact, infectious viruses,** known as Dane particles. **HBeAg**: Sera of infected individuals with active disease contains HBeAg, a component of the viral capsid. There are other antigens found in HBV infection but they are not found in serum at high concentrations. These antigens are described later.

DNA Viruses

Hepatitis B Virus

ss DNA

protein associated with DNA

complete (-) strand

incomplete (+) strand

DNA dependent DNA polymerase

1. After entry into the cell's nucleus the virus under goes genome maturation where the + strand of the DNA is completed using the - strand as a template.

2. A full length +RNA is transcribed from the - DNA strand and is called the pre-genome.

DNA dependent RNA polymerase

RNA pre-genome (+)

4. The DNA dependent DNA polymerase transcribes a partial +DNA strand from the - DNA strand to complete the replication.

DNA dependent DNA polymerase

RNA dependent DNA polymerase

3. The pre-genome, reverse transcriptase (RNA dependent DNA polymerase), DNA dependent DNA polymerase and the terminal proteins are all packaged inside a nucleocapsid. The reverse transcriptase then transcribes the -DNA strand while destroying the RNA pre-genome.

Nucleocapsid

Hepatitis B Diseases

Transmission of hepatitis B is primarily through **blood**. Only very small amounts of blood are required for infection, i.e. accidental needle sticks can produce infection. IV drug users who share needles are particularly at risk for infection.

Mother to child transmission may occur during passage through the birth canal or during breast feeding.

Sexually transmitted disease. Homosexual men traditionally have had a very high rate of infection which has decreased with the introduction of the HBV vaccine and safe(r) sexual practices.

After entering the blood, the organism reaches the liver and enters hepatocytes.

Acute hepatitis is clinically similar to hepatitis A, except the symptoms last longer and can be more severe.

Incubation period: **very long**, 50-180 days (2-6 months).

Symptoms: fever, rash, arthritis, headache, nausea and jaundice are common symptoms although infection can be asymptomatic. As HBsAg disappears so do the symptoms. The duration of the disease is from 8-10 weeks.

Acute hepatitis can be **self limiting** (asymptomatic or with jaundice lasting 4-5 weeks) or develop into **chronic hepatitis**. The disease may also progress to **fulminant hepatitis** which is a rapidly progressive disease which can be fatal in 7-10 days. In fulminant hepatitis there is extensive liver damage and mortality can be up to 80%.

Chronic hepatitis is defined as the persistence of HBsAg in the sera for 6 months or longer and occurs in about 10% of those with hepatitis B infection. These people can transmit the disease.

Two types of chronic hepatitis:

1. **Chronic persistent** is characterized by minimal necrosis usually associated with a favorable outcome.

2. **Chronic active** is more severe; characterized by progressive necrosis of hepatocytes, destruction of the liver architecture, and fibrosis.

The microscopic hallmarks of chronic active hepatitis are **piecemeal necrosis** and bridging necrosis. This destruction can lead to **cirrhosis** and/or primary **hepatocellular carcinoma** (HCC). HCC is the most common cancer in the world and people with chronic active hepatitis have 200 times the risk.

Treatment with interferon for chronic infection; no treatment for acute.

Prevention: a recombinant **vaccine** containing HBsAg is now a routine vaccination of childhood. It is also recommended for all health professionals and other high risk groups.

Hepatitis B immunoglobulin (HBIG) consists of HBsAb IgG and is a passive vaccine given to individuals exposed to the virus (needle stick injury, sexual contact with a carrier).

DNA Viruses

Hepatitis B Virus

The Microbiology Companion, Topf and Faubel ©1997

HBV Timeline: serologic tests for HBV antigens and their antibodies provide useful information about the presence and stage of infection. The order and duration of the appearance of the antigens and antibodies make excellent test questions.

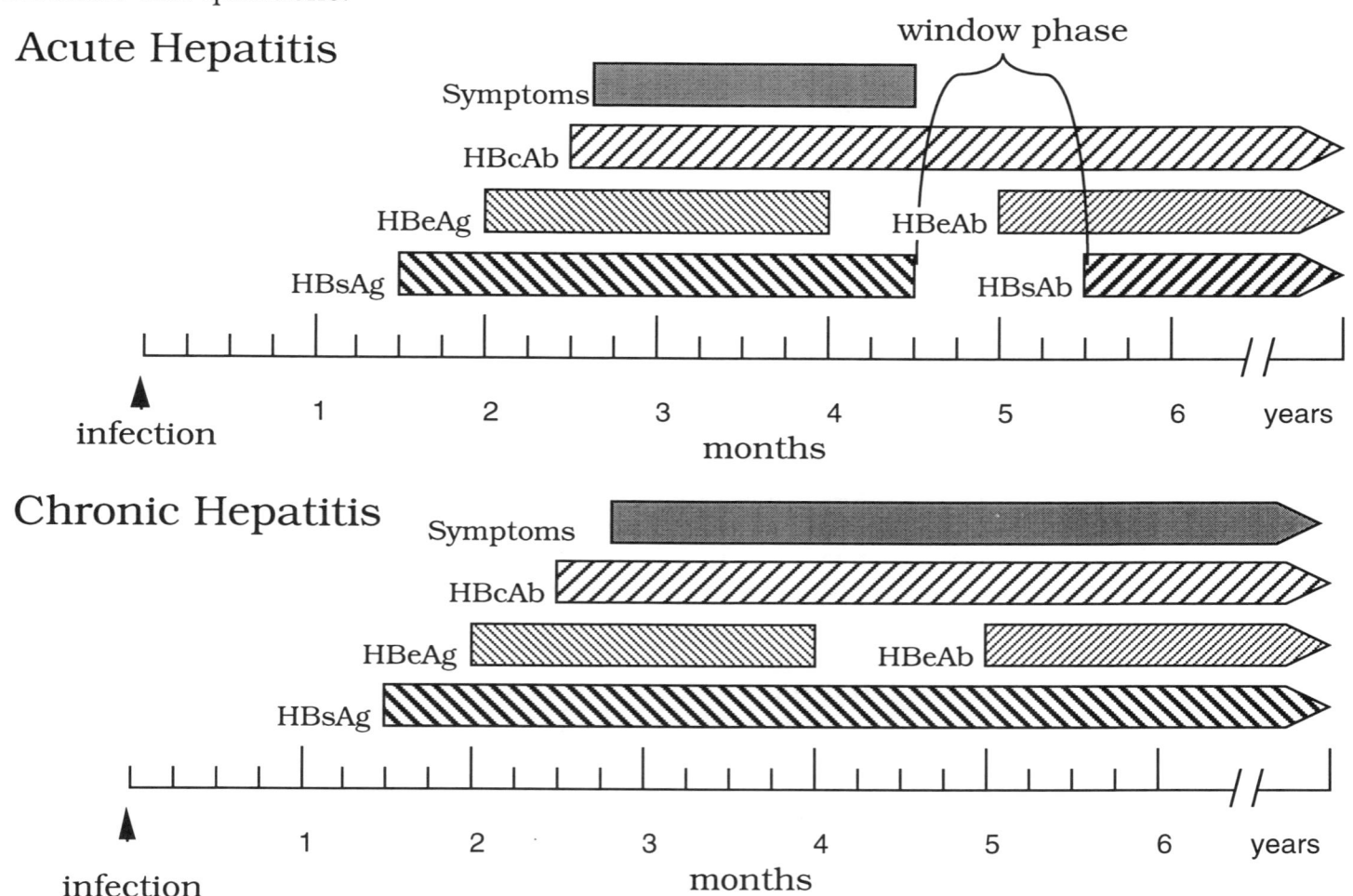

Summary of HBV Antigens and Antibodies

	antigen	antibody
SURFACE	**HBsAg** Present during incubation, prodrome and acute infection. Most important lab test for detection of disease. It is the <u>first antigen</u> to be detected in serologic tests (appears within one month of exposure to HBV). Presence for over 6 months indicates chronic disease and carrier state. HBsAg is given as a routine childhood vaccine and to select populations which are at risk. HBsAg for the vaccine is either purified from infected individuals or genetically produced in yeast.	**HBsAb** Appears at about 6 months after initial infection; indicates recovery and provides <u>protection against reinfection</u>. Persons given HBV (active) vaccine will have antibodies to the surface antigen only (HBsAb). Hepatitis B immunoglobulin (HBIG) contains HBsAb and is used for immediate passive immunization (i.e. after a needle stick).
	A **window period** of several weeks exists between the disappearance of surface antigen and the appearance of antibody to the surface antigen when neither can be detected in the serum.	
CORE	**HBcAg** A serologic test for this antigen does not exist.	**HBcAb** <u>First antibody</u> to appear in serologic tests (appears about 2 months after exposure). Is positive during the window phase. Will only be present for 6 months. Presence indicates active disease, not found in chronic infection.
E	**HBeAg** Present during incubation, prodrome and acute infection; appears 2-4 weeks after HBsAg (about 1 and 1/2 months after exposure). Presence indicates transmissibility.	**HBeAb** Second antibody to appear in serologic tests (appears about 5 months after exposure). Presence indicates low transmissibility. Presence indicates good prognosis.

DNA Viruses

DNA Viruses

Hepatitis Delta Virus (HDV)

Infection with Hepatitis D can complicate HBV infection

Small SS circular RNA	HDV virus is a **defective virus** because it is does not contain the genetic instructions for a complete infectious virus. It must depend on another virus (in this case HBV) to code for its envelope.
Envelope is a hepatitis B envelope, HBsAg.	
Genome codes for a single protein the delta antigen.	Transmission of HDV occurs through blood transfusions and needle sticks.

Two types of HDV infections can occur:

Co-infection:

- Infection with HBV and HDV occurs simultaneously as in a blood transfusion which harbors both viruses.
- The infection is indistinguishable from the hepatitis caused by HBV alone.
- The HBV and HDV infection usually resolves.
- There is no increase in the incidence of chronic hepatitis.

Super-infection:

- Infection with HDV occurs in a person who already has chronic HBV.
- The additional virus increases the severity of the hepatitis and chronic HDV infection usually occurs.
- The incidence of severe liver damage (chronic cirrhosis) is increased.
- The mortality rate is increased due to the extensive liver damage.
- Think: super-infection is super bad.

DNA Viruses

DNA Viruses

Poxviridae

- Most complex virus.
- DNA codes for over 100 proteins.
- Only virus with a double layered membrane.
- Only DNA virus which does not enter the nucleus.

characteristics	life cycle	notes
enveloped (a double membrane) brick shaped complex nucleocapsid DS linear DNA one serotype **DNA dependent RNA polymerase** is carried with the virus because it **replicates in the cytoplasm**.	**All steps of replication occur in the <u>cell cytoplasm</u>.** 1. Penetration of the host cell can occur by phagocytosis or fusion with the cell membrane. The cell then uncoats to release the viral core which is still enclosed in the inner membrane. 2. **DNA dependent RNA polymerase** synthesizes early mRNA which is extruded out of the viral core and translated by host cell ribosomes into early proteins. 3. Early proteins digest the viral coat releasing the viral DNA. 4. A DNA dependent DNA polymerase carries out replication of the viral DNA. 5. Late mRNAs and late proteins (i.e. structural proteins) are synthesized. 6. Virions are assembled. 7. Double membrane is acquired (in part through the golgi apparatus) and the complete virus is released.	The envelope is not acquired by budding through the cell membrane (see step 7). Virions with and without the envelope are infectious. Induces cell proliferation which accounts for the pox lesions. The lesion heals when the infected cells die. Most viruses use host cell DNA dependent RNA polymerase found in the nucleus to produce mRNA, but since pox virus does not enter the nucleus it must cary its own RNA polymerase.

Poxvirus

DNA Viruses

Poxvirus

The Microbiology Companion, Topf and Faubel ©1997

104

Poxvirus	diseases	notes
Smallpox (variola) virus	**Smallpox** is the only disease that has been completely eradicated throughout the world. The last case of natural infection was in Somalia in 1977. In 1967, the World Health Organization (WHO) initiated a campaign to eliminate smallpox. This endeavor included worldwide vaccinations and quarantines of infected individuals. Eradication was possible for the following reasons: • one serotype of the smallpox virus • the disease only infects humans, no animal reservoir • no asymptomatic or latent infections occur Two varieties of the disease: Variola major: mortality rate of 25-50% Variola minor: mortality rate less than 1% **Transmission** is by respiratory aerosol or direct contact. After a 1-2 week incubation, a prodrome of fever, chills and malaise occurs. This signals a generalized viremia as the virus spreads throughout the body. 3 to 4 days later a rash appears on face and spreads to the entire body. The rash progresses through set stages of development: 1. raised firm papules 2. fluid filled vesicles 3. rupture of vesicles 4. crusting and healing of lesions	**Vaccine** contains **live** vaccinia virus (see below). The virus is purified from skin lesions of infected calves. Inoculation of the vaccine is by intradermal injection. The formation of a lesion indicates successful inoculation. The vaccine is no longer routinely administered. Military personnel and laboratory workers with possible exposure are given the vaccine. The last smallpox virus is held in two freezers. One in Atlanta at the CDC and the other in Moscow. These last samples are set for termination in spring 1995. It will be the only species ever intensionally wiped out by people.
Vaccinia	**Nonpathogenic** The origin of vaccinia virus is unknown, but it may be a recombinant of cow and human smallpox or a type of horse poxvirus. Live virus is used to immunize against smallpox (described above).	There is interest in using vaccinia to carry other viral antigens and use it as a vector for other immunizations.
Molluscum contagiosum virus	**Cutaneus infection** which forms benign wart-like tumors of the skin without systemic symptoms. **Transmission** occurs by direct contact with skin lesions. After a 1-2 month incubation, small papular, pale lesions appear on the skin. The lesions are **self-limited** and will disappear after 2 months to a year.	

Notes

DNA Viruses

Herpesviridae

The most notable characteristic of Herpesviridae is the ability to cause latent infections by inserting their DNA into the host cell genome. Then at some later date, often in response to the stress of illness, the viral DNA leaves the host genome and produces an infection that can be quite different from the primary disease.

Herpesviridae have <u>DS linear DNA</u> with an <u>icosahedral capsid</u> and <u>lipid envelope</u>. The envelope is unique in that it is acquired when the virus buds through the **nuclear membrane**.

life cycle

1. Entry into the host cell by endocytosis and release of the viral capsid into the cytoplasm.
2. Viral DNA enters the nucleus.
3. In the nucleus, viral mRNA is created with host cell polymerases. The first genes to be transcribed are called the alpha genes which code for proteins called the immediate early proteins (also called the alpha polypeptides). The immediate early proteins are required to transcribe any other mRNA.
4. Using the immediate early proteins, the rest of the DNA is transcribed into mRNA. There are three classes of genes and proteins:

alpha genes	code immediate early proteins alpha polypeptides	needed to transcribe viral mRNA
beta genes	code early proteins beta polypeptides	used for DNA replication
gamma genes	code late proteins gamma polypeptides	structural proteins

5. Progeny DNA is produced by a viral DNA polymerase (blocked by acyclovir, an antiviral drug).
6. Assembly: capsid proteins enter the nucleus and combine with progeny DNA.
7. Membrane proteins are inserted into the nuclear membrane.
8. Viral membrane is acquired by budding through the **nuclear** membrane.
9. A single host cell can produce 10,000 virions.

Herpesviridae: 8 different viruses

virus	latent virus remains in	primary infection	recurrent infection
Herpes simplex virus-I	sensory ganglia of the head	gingivostomatitis keratoconjunctivitis herpetic whitlow meningoencephalitis	herpes labialis (cold sores) keratoconjunctivitis encephalitis
Herpes simplex virus-II	sensory ganglia of lumbar and sacral regions	genital infections neonatal infections	genital infections
Varicella-zoster virus	sensory ganglia of the head or chest	varicella (chickenpox)	zoster (shingles)
Cytomegalovirus	leukocytes	heterophile (-) mononucleosis congenital cytomegalic inclusion disease of the fetus CMV retinitis	reactivation in pregnant women may result in congenital cytomegalic inclusion disease of the fetus
Epstein-Barr virus	B-cells	heterophile (+) mononucleosis hepatitis	Burkitt's lymphoma nasopharyngeal carcinoma
Human herpesvirus 6	unknown	roseola (sixth disease) febrile seizures	possibly, interstitial pneumonia
Human herpesvirus 7	unknown	possibly roseola	unknown
Human herpesvirus 8	unknown	unknown	possibly, Kaposi's sarcoma

DNA Viruses

virus	diseases	notes
HSV-I Herpes simplex virus I ♦ DS linear DNA ♦ icosahedral nucleocapsid ♦ lipoprotein envelope	**In general HSV-I causes oral infections.** Primary infection is generally before puberty. **Transmission** is by contact with vesical secretions. Latent infection occurs in the sensory ganglia of the head. **Gingivostomatitis** is a primary infection that is symptomatic in 15%. Infection is characterized by inflammation of the gingival and oral mucosa. Painful **vesicles** and **ulcerations** may appear. Lasts **5-12 days**, virus is shed for only the first 4 days. Virus becomes latent in the trigeminal ganglion and recurrences are less severe than the primary infection. Preschool aged children are usually affected. **Herpes labialis** (cold sores, fever blisters) is a reactivation of infection that may **recur** throughout the life of an infected individual. Sores usually appear around the mouth or on the lips and last about a week. Reactivation is unilateral. Primary infection occurs <u>inside</u> the mouth and recurrent infection occurs <u>outside</u> the mouth. **Herpetic whitlow** is an infection of the skin and nail area which forms pustules. Can cause loss of skin resulting in dehydration and 2° bacterial infection. Infection usually occurs through a small break in the skin or in an area of eczema. Health care workers are at risk for this infection because they touch herpetic lesions. **Keratoconjunctivitis** is an infection of the cornea and conjunctiva of the eye. Recurrence may occur due to reactivation of latent infection. Multiple infections from recurrence causes scarring which may lead to blindness. **Encephalitis** usually affects the temporal lobe and has a high mortality rate. The pathogenesis is not well understood.	Only 1% of all primary infections are symptomatic. 25% of all people infected with HSV I or II will experience recurrent infections. The infection is widespread and the majority of people have been infected by adulthood. **Herpetic encephalitis** is the most common cause of fatal sporadic (as opposed to epidemic) encephalitis in the U.S. (250 deaths per year). Reactivation of latent infection has been associated with: menstruation, emotional stress, sunlight and fever. Cell mediated immunity is required to overcome a 1° or recurrent infection. Individuals with impaired cell mediated immunity are susceptible to chronic disseminated disease. Treatment options include acyclovir and famciclovir; foscarnet is used if acyclovir resistant.

virus	diseases	notes

HSV-II
 	
Herpes simplex virus II
 	
♦ DS linear DNA
♦ icosahedral nucleocapsid
♦ lipoprotein envelope
	In general HSV-II causes genital tract infections.
	
	1° infection is generally after puberty.
	
	Transmission is by contact with vesical secretions.
	
	Latent infection occurs in the sensory ganglia of lumbar and sacral regions.
	
	Genital herpes is a sexually transmitted disease. Transmission is by contact with genital lesions.
	
	   Primary infection occurs after an incubation period of 5 days when small papules form in the genital area. The papules progress to vesicles and then to pustules which burst to leave painful ulcers. May last a month.
	
	   The lesions are multiple (20-30) and bilateral. External and internal (cervix, urethra) structures are affected.
	
	   Systemic symptoms of fever, malaise, and headache may be present. Inguinal lymph nodes are often enlarged and tender.
	
	   Asymptomatic infections occur in men and women and are an important source of infection.
	
	   Recurrent infection is shorter and milder than the primary infection. Recurrence is often preceded by a burning sensation in the genital area. Lesions may be itchy and painful. Usually there are no systemic symptoms. Virus can be shed between recurrences (while patient is asymptomatic).
	
	Neonatal herpes is acquired during passage through the birth canal of an infected mother. If lesions are present just prior to delivery, the child should be delivered by C-section.
	
	   HSV may disseminate to internal organs or the CNS. The mortality rate is very high (above 60% for disseminated disease, 15% for CNS disease). If the infant survives, neurological and optical sequelae are common. The risk is greater if the 1° disease occurs during the first trimester.
	
	Cervical carcinoma: higher titers of HSV-II antibodies have been found in women with cervical cancer than in controls. However, the significance of the association between HSV-II and cancer is uncertain and it appears that **human papilloma virus** plays a more important role in the development of cervical cancer.
		Tzanck smear is used in the diagnosis of HSV-I and II infections. A smear from a lesion is stained and checked for the presence of **intranuclear inclusion bodies or multinucleated giant cells.**
		
		Treatment:
		
		   Acyclovir is used to shorten the duration and reduce the infectivity of genital herpes, but it has no effect on the virus in the latent state.
		
		   Vidarabine (an anti-viral drug which works the same way as acyclovir) and acyclovir are used in the treatment of neonatal herpes with a 10% reduction in mortality.
		
		HSV-II is occasionally implicated in oral lesions.
		
		HSV-I is increasingly found to be a cause of genital herpes.

DNA Viruses

DNA Viruses

Varicella-zoster virus (VZV)

- DS linear DNA
- icosahedral nucleocapsid
- lipoprotein envelope

primary infection	recurrent infection	diagnosis and treatment
Varicella (chickenpox) is primarily a disease of children. 90% have antibodies to the virus by age 10. **Transmission** occurs via respiratory droplets, most commonly in the winter and spring. The incubation period is **2 weeks**, during which the virus becomes disseminated via the blood. Fever and headache precede the rash. A **maculopapular rash** appears on the trunk and spreads outward. The rash is pruritic (itchy) and progresses to pustules which rupture and form scabs. The rash appears in crops, i.e. multiple eruptions occur over time so that patients have lesions in all stages at one time. This differentiates varicella from smallpox in which all of the lesions are the same age. Symptoms last about 2 weeks. Immunity to varicella is lifelong. The disease is mild is children, but can be severe in adults and the immunocompromised. Treatment is reserved for pregnant women and the immunocompromised; use acyclovir. **Pneumonia** is the most common complication of infection in adults. The morbidity and mortality of pneumonia is increased in pregnant women. **Meningoencephalitis** is a very rare and highly fatal complication that occurs in all age groups. **Reye's syndrome** is characterized by encephalopathy and liver damage. Associated with VZV or Influenza B. Occurs more often in children given aspirin. **Congenital varicella syndrome**: Primary maternal VZV infection in the first trimester can cause severe CNS defects in the fetus.	Latent infection occurs in the cells of sensory ganglia, often the dorsal root ganglia near the spinal cord. **Zoster** (shingles) is a reactivation of a latent infection. May present with severe burning skin pain prior to the appearance of vesicles. If pain occurs in the chest, it can be confused with an acute MI. It can result from immunosuppression, trauma or other stress. Painful vesicles are limited to the distribution of the infected nerve (dermatomal distribution, can be unilateral). The lesions often appear on the head or follow the course of an intercostal nerve. Immunity is not protective and zoster may recur numerous times. Patients with zoster can transmit chicken pox. **Post-herpetic neuralgia** is a pain syndrome which is a sequela of zoster. Treatment with an anti-viral (e.g. acyclovir) speeds healing and reduces the severity and duration of post-herpetic neuralgia .	Has only one serotype. VZV can be **grown in a cell culture** of human fibroblasts. **Diagnosis** is usually made on clinical grounds. **Tzanck** smear can be used in the diagnosis of VZV infections. A smear from a lesion is stained and checked for the presence of **intranuclear inclusion bodies or multinucleated giant cells.** **Prevention:** **Acyclovir** is used as prophylaxis for immunocompromised persons exposed to the virus. **VZV immune globulin** is also available for prophylaxis. **Vaccine** contains live attenuated virus and was approved for use in 1995. The vaccine may be given any time after 1 year of age.

DNA Viruses

Cytomegalovirus (CMV)

- DS linear DNA
- icosahedral nucleocapsid
- lipoprotein envelope

prenatal disease	adult disease	disease in immunocompromised	diagnosis and treatment
Cytomegalic inclusion disease occurs in the **fetus** by <u>transmission of the virus across the placenta</u>. Can result in the death of the fetus. CMV is the **most common congenital viral infection**. 1% of newborns are infected and of those 10% show abnormalities which include: microcephaly (small head), hepatosplenomegaly, deafness, mental retardation and jaundice. The infant may appear normal at birth and not show abnormalities until later in life. Fetal infection may occur after a primary infection of the mother, or, more often, from reactivation of a latent infection. A primary infection causes more severe abnormalities. The maternal infection is most commonly asymptomatic.	**Heterophile-negative mononucleosis** is an adult infection clinically similar to the mononucleosis caused by EBV but without the sore throat. It is characterized by the symptoms of fever, chills, headache and fatigue. There is no agglutination of sheep RBCs by the patients sera (heterophile negative). 80% of adults have antibodies to the virus and almost all infections are asymptomatic.	Persons on immunosuppressive drugs after **organ transplant** are especially prone to CMV infection. CMV is sometimes transmitted in a donor organ. Cell mediated immunity is more important than humoral immunity in the body's defense against CMV. **Pneumonia** is most common is transplant recipients, especially bone marrow. **CMV retinitis** attacks the eyes of people with AIDS. It causes progressive white-yellow plaques to form on the retina. It leads to blindness. CMV is an immunosuppressive infection. It decreases CD+4 T-cells. May have a role in enhancing/promoting HIV infection.	Latent infection occurs in leukocytes. **Transmission** of CMV can occur across the placenta, during passage through the birth canal. or from breast milk, saliva, urine, cervical secretions and semen. Infected cells are enlarged (cytomegalic) and have both cytoplasmic and intranuclear inclusions. CMV can be **grown in a cell culture** of human embryonic fibroblasts. Only one serotype. May be excreted in urine and saliva for years after initial infection. Treatment is with IV ganciclovir or foscarnet.

Notes

DNA Viruses

Epstein Barr Virus (EBV)

- DS linear DNA
- icosahedral nucleocapsid
- lipoprotein envelope

primary disease	malignancies and associations	notes
Heterophile-positive infectious mononucleosis is a disease of young adults which is transmitted by **saliva**.	Latent infection occurs in B-cells.	90% of adults in the U.S. have antibodies to the virus.

primary disease

Heterophile-positive infectious mononucleosis is a disease of young adults which is transmitted by **saliva**.

Infection is characterized by fever, chills, sweats, fatigue, **severe sore throat**, enlarged lymph nodes, splenomegaly and/or mild hepatitis.

The virus transforms infected B-lymphocytes such that there is a polyclonal expansion. This is essentially a malignant transformation except the transformation ends when the virus and transformed cells are eradicated by the immune system.

Diagnosis is by observation of atypical, **large lymphocytes** (T-cells) in a peripheral blood smear. Viral antigens on infected B-cells induce T-cell proliferation and differentiation into atypical lymphs. The ≠T-cells cause lymphadenopathy, swollen tonsils (leading to the sore throat) and destruction of the infected B-cells.

> **Heterophile antibodies** are antibodies that bind antigens other than the antigens that induced them.

> EBV induces antibodies which bind to sheep RBCs. The antigen which induces these antibodies are normal host cell membrane proteins which have been altered by the virus. These antibodies do not react with EBV or any of its proteins.

Heterophile antibodies are detected by using the patients serum to agglutinate sheep RBCs.

malignancies and associations

Latent infection occurs in B-cells.

Burkitt's lymphoma is a lymph node malignancy of the head and neck.

> The disease occurs worldwide, but is only associated with EBV in Africa where this type of lymphoma is the #1 childhood cancer. This geographic distribution suggests that a cofactor is probably involved in producing malignancy. Since malaria is common in the same areas as Burkitt's lymphoma, it is hypothesized that malaria may be the cofactor. Malaria lowers host immunity which could allow B-cell proliferation to go unchecked.

> (8,14) or (2,22) chromosomal translocation are seen in practically all malignant cells of Burkitt's lymphoma. These translocations move the cellular oncogene *myc* next to a gene responsible for antibody production. Since B-cells will transcribe the gene for antibodies at a high rate, the misplaced oncogene will be overproduced.

Nasopharyngeal carcinoma occurs in south China and associated with EBV.

Chronic fatigue syndrome is a recently characterized syndrome of unexplained fatigue with fever, myalgia, and headaches. Typically it strikes after an acute infection. High titers of EBV antigens have been found in many patients.

notes

90% of adults in the U.S. have antibodies to the virus.

Viral capsid antigen (VCA) is used in diagnostic tests.

> Anti-VCA **IgM** is the first antibody produced and is useful to diagnose acute infection.

> Anti-VCA **IgG** follows next and persists for life; it is used to diagnose past infection and immunity.

Viral membrane antigen (VMA) induces antibodies protective against infection.

Immunity to the virus is lifelong.

The virus can cause a latent infection in **B-cells**. This infection allows the lymphocytes to proliferate continuously.

Human HerpesvirusHuman Herpesvirus (HHV)

virus	primary infection	secondary infection	characteristics
Human herpesvirus–6	**roseola**, also known as exanthem subitum or sixth disease, is a common childhood illness characterized by high fever followed by the onset of a rose colored rash. **acute febrile illness** may occur without the development of rash. **febrile seizures** in infancy are associated with HHV-6 infection. **infectious-mononucleosis-like illness** can occur in adults.	Specific illness related to reactivation of HHV-6 infection have yet to be proven. Reactivation is probably asymptomatic in the immunocompetent, and has been associated with the following in the immunocompromised **febrile illness** **interstitial pneumonitis** **bone marrow suppression** **graft rejection** in transplant patients. HHV-6 infection is also associated with certain malignancies: **Hodgkin's disease** **non-Hodgkin's lymphoma** **leukemia**	90% of children have evidence of infection by age 3. Mechanism of transmission is probably from saliva Can co-infect T-lymphocytes with HIV; unknown if there is consequential interaction between these viruses. Causes malignant transformation of certain cells in cell culture.
Human herpesvirus–7	Although HHV-7 has been associated with some cases of **roseola**, the importance of HHV-7 as a cause of human disease is yet unknown.	unknown	Like HHV-6, a large percentage of the population is infected at an early age. Probably > 85%. Mechanism of transmission is probably through saliva. First isolated from CD4+ T cells of a healthy host.

DNA Viruses

virus	primary infection	secondary infection	characteristics
Human herpesvirus–8 Also known as KS-associated herpesvirus or KSHV-1	unknown	HHV-8 is proposed to be the etiologic agent of **Kaposi's sarcoma** (KS). Kaposi's sarcoma is the most common malignant neoplasm seen in people with AIDS and is an AIDS defining illness. The disease most commonly effects the skin, oral cavity, lungs and GI tract. HHV-8 has been isolated from KS lesions in patients with and without HIV infection. Prior treatment with foscarnet (but not other anti-herpesvirus drugs) has been associated with a decreased incidence of KS in people with AIDS. HHV-8 is also thought to cause **interstitial pneumonia**.	

Papovaviridae

- DS DNA, circular
- icosahedral nucleocapsid
- no envelope

The name comes from the first two letters of the original three viruses in this family:

papillomaviruses

mouse **po**lyomavirus

simian **va**cuolating virus

Life cycle: the structure and replication of these virus has been studied in depth and is used as a model for eukaryotic DNA expression and replication.

1. Penetration, uncoating, and transfer of the DNA into the nucleus.
2. Transcription of viral DNA by host DNA dependent RNA polymerase II into mRNA.
3. The first genes to be transcribed are the T-antigens (*Tag* and *tag* in SV40, notice that these two antigens are distinguished by the case of the first letter T vs. t).
4. A single mRNA produces both Tag and tag. The Tag mRNA requires the splicing out of an intron which results in a larger and protein.

The T-antigen DNA has two stop codons. These terminate protein synthesis at the ribosome.

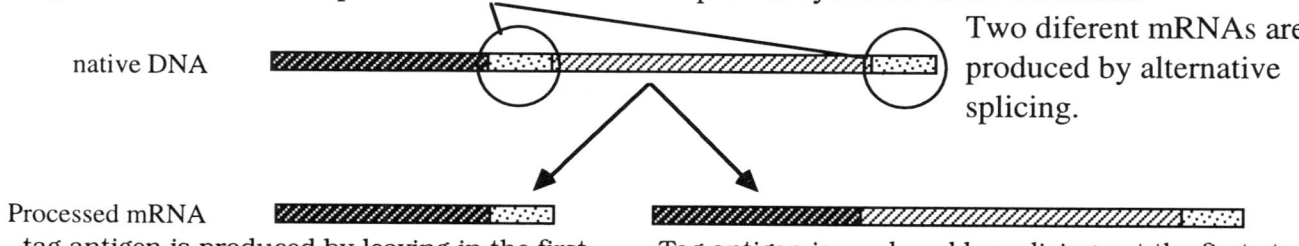

native DNA

Two diferent mRNAs are produced by alternative splicing.

Processed mRNA

tag antigen is produced by leaving in the first stop codonwhich creates a short protein.

Tag antigen is produced by splicing out the first stop codon which allows a much larger protein to be translated.

5. Tag is required for late gene expression and for DNA replication.
6. Capsid proteins are translated and the virus is assembled in the nucleus.
7. Progeny viruses are released by cell lysis.

Papovaviridae

The Microbiology Companion, Topf and Faubel ©1997

polyoma virus	viral characteristics	pathogenesis
Mouse polyoma virus DS circular DNA icosahedral capsid no envelope	**Mice** are the natural host. In adult mice, the infection is asymptomatic. When injected into newborn mice, malignant tumors develop all throughout their little bodies.	Two cycles occur within host cells: **Lytic cycle** is essentially the standard viral infection in which the host cell produces many progeny viruses and then dies. **Nonpermissive cycle** occurs when progeny are not produced and the cell is transformed in to a malignant cell. In this cycle the viral DNA is incorporated into the host cell genome and only the **early genes** (tag) are transcribed.
SV40 **Simian vacuolating virus** DS circular DNA icosahedral capsid no envelope	**Monkeys** are the natural host. Does not cause malignant transformation in monkeys but it can transform rodent and human cell cultures. The Salk and Sabin polio vaccines were grown in monkey kidney cells and were contaminated with SV40. This resulted in the inoculation of millions with live SV40 virus. The effect of this contamination is still unknown.	The genes of mouse polyoma virus and SV40 code for a group of proteins called the **T antigens**. The **polyoma** virus has 3 T antigens. The **SV40** virus has only 2 T antigens. The largest **T antigen is both necessary and sufficient** to cause malignant transformation. It alters the regulation of host cell DNA replication. The large T antigen is a protein kinase which binds to DNA and stimulates its synthesis.
Human polyomavirus JC virus DS circular DNA icosahedral capsid no envelope	**Progressive multifocal leukoencephalopathy (PML)** is a rare demyelinating disease characterized by memory loss, aphasia and loss of coordination. Immunocompromised persons are susceptible. Produces brain tumors in infant hamsters.	JC virus is ubiquitous and most people are seropositive by adolescence. The virus was first isolated in 1971 from the brain of a patient with Hodgkin's lymphoma. The virus was named for the initials of this patient.
Human polyomavirus BK virus DS circular DNA icosahedral capsid no envelope	Found in 10-40% of renal transplant patients. Causes asymptomatic or mild respiratory infections in children.	BK virus is ubiquitous and most people are seropositive at a young age. The virus was first isolated from the urine of a renal transplant patient. The virus was named for the initials of this patient.

Notes

DNA Viruses

Human papilloma virus

♦ DS circular DNA
♦ icosahedral capsid
♦ no envelope

pathogen	diseases	important notes
Human papilloma virus (HPV) DS circular DNA icosahedral capsid no envelope	53 different genotypes cause cutaneous warts and a variety of other squamous lesions throughout the body. Usually the lesions are benign and regress spontaneously. **Common warts** are caused by HPV-1, 2, 3, 4, 10, 26, 28. Occur in the thick keratinized skin of the palms or feet. Warts will spontaneously regress. **Condylomata acuminata (genital warts)** is one of the most common **sexually transmitted disease** (infects 20-60% of adult women). Caused by HPV-6, 11, 30. Lesions regress spontaneously but can recur. **Cervical carcinoma** is associated with HPV-16, 18, 31. HPV DNA is integrated into host cell DNA and plays an unclear role in transformation. **Epidermodysplasia verruciformis** is an inherited disease associated with inadequate cellular mediated immunity and infections with HPV-5 and 8. It is characterized by warts and macular lesions. About 1/3 become <u>malignant</u>.	The various genotypes are determined by comparing DNA. A genotype must be more than 50% different than any other genotype to be considered unique. Do not confuse genotype with serotype. Serotypes are based on antigenic not genetic differences. Malignant cells harbor viral DNA which only translates **early genes**. Two genes which are carcinogenic are **E6** and **E7**. They bind and block the proteins coded for by the tumor suppressor genes p53 and retinoblastoma gene Has **never been grown in cell culture**. Diagnoses is clinical, koilocytosis is a finding on Pap smears which indicates HPV infection. Acetic acid will reveal subclinical warts by turning the lesion white.

Adenoviridae

- DS linear DNA
- icosahedral capsid
- no envelope

pathogen	structure	diseases	life cycle and important notes
Adenovirus	Over 40 serotypes affect humans. **Fibers** project from each penton (the corner molecules) of the icosahedral capsid. The fiber is a **hemagglutinin** which binds the surface of RBCs; serotypes can be determined by hemagglutination inhibition test. Purified fibers are cytotoxic. The linear DNA has a protein covalently bound to the 5' end which stabilizes it and promotes DNA replication.	Transmission is person to person through respiratory droplets, contact with tears and fecal-oral. **Respiratory infections** (serotypes 2, 1, 7, 3, 5 in order of frequency) are rare in adults, but cause upper and lower respiratory infections in **children**. Military recruits get epidemic **acute respiratory disease** (serotypes 4, 7) which can cause pneumonia and can be fatal. **Epidemic keratoconjunctivitis** (3, 7) associated with dusty environments and hand-eye contact. **Gastroenteritis** (40, 41) occurs in young children. **Hemorrhagic cystitis** (11) causes hematuria (bloody urine) and dysuria (painful urination). Rodents injected with the virus have malignant transformation of cells (12, 18, 31). **No adenoviral DNA has ever been found in any human cancer.**	**Life cycle** 1. Attachment to host cells via capsid fiber. 2. Entry and uncoating. 3. Transport of DNA to nucleus. 4. Expression of early genes. These proteins are required for DNA replication. One of the proteins is a DNA polymerase which allows replication of the viral DNA. 5. The reproduced DNA is then used as a template to produce the late mRNA. The late mRNA is transcribed as a single RNA which is processed to produce all of the various structural proteins. 6. Virions are assembled in the nucleus and released by cell lysis. The virus can be recovered from the adenoids and tonsils of asymptomatic individuals. The carriers can shed virus asymptomatically for years. Infection produces long lasting type specific immunity. The illness occurs primarily in children. **Vaccine** contains live (nonattenuated) virus and is used only in military recruits (4, 7).

The Microbiology Companion, Topf and Faubel ©1997

DNA Viruses

DNA Viruses

Parvoviridae

- ♦ **single stranded** linear DNA
- ♦ icosahedral capsid
- ♦ no envelope

pathogen	characteristics	diseases
B19	Smallest icosahedral virus. DNA may code for only 3 proteins. Very durable virus: able to survive heating at 56°C (132°F) for hours. Resistant to both chloroform and ether. Remain viable even after years of storage. Transmission is by respiratory droplets.	**Erythema infectiosum** (Fifth disease) is a mild childhood disease characterized by a facial rash which looks like <u>slapped cheeks</u>. Often asymptomatic. **Erythroid precursor** cells are a favored target of B19 virus. The virus causes the lysis of these cells decreasing the supply of new RBCs. **Aplastic crisis** may occur in persons with chronic hemolytic anemia (sickle cell anemia, congenital spherocytosis, etc.). RBCs have a shorter life span in persons with hemolytic anemia and infection with B19 compounds the problem by increasing the destruction of RBCs. Aplastic crisis occurs when the body can't keep up with the demand for new RBCs. **Chronic anemia** may develop in immunodeficient persons. **Spontaneous abortion** has been attributed to B19.

Think: parvovirus = part-of-a-virus: only DNA virus with single stranded DNA.

DNA Viruses

Summary of Enveloped RNA Viruses

family	genus/group	virus/common name	structure	notes
Orthomyxoviridae	Influenza virus	Influenza viruses A, B, C causes the flu	segmented SS (-) RNA helical capsid	The RNA genome is replicated in the nucleus.
Paramyxoviridae	Morbillivirus Paramyxovirus Pneumovirus	Measles Parainfluenza -- causes croup Mumps Respiratory syncytial virus	non-segmented SS (-) RNA helical capsid	The RNA genome is replicated in the cytoplasm.
Togaviridae	Rubivirus 　no insect vector Alphavirus 　insect vector	Rubella virus Eastern equine encephalitis virus Western equine encephalitis virus	non-segmented SS (+) RNA icosahedral capsid	The (+) RNA allows these genomes to infect cells without any viral proteins. They are infectious genomes.
Flaviviridae	Flavivirus	Saint Louis encephalitis virus Yellow Fever virus Dengue Fever virus Hepatitis C *	non-segmented SS (+) RNA icosahedral capsid	Infectious genomes use their (+) RNA as mRNA right after uncoating.
Rhabdovirus	Lyssavirus	Rabies Virus	non-segmented SS (-) RNA bullet-shaped capsid	
Bunyaviridae	Bunyavirus	Hantaviruses	three segments SS (-) RNA helical capsid	
Filoviridae		Ebola virus Marburg virus	non-segmented SS (-) RNA helical capsid	

* Hepatitis C is Flavivirus-like, it has not been officially classified yet.

Orthomyxovirus

- ◆ segmented SS (-) RNA
- ◆ helical capsid
- ◆ envelope with H and N

pathogen	structure	life cycle
Orthomyxovirus Influenza A Influenza B Influenza C	**Segmented SS (-) RNA** The RNA genome is divided into 8 (C has only 7) pieces, which codes for a total of ten proteins, each segment codes for one protein and two segments use alternative splicing to generate two proteins. **Helical nucleocapsid** Three antigenically distinct ribonucleopro-teins divide the virus into 3 groups: A, B, and C. The ribonucleoproteins (NP) are **group specific antigens**. Antibodies against the ribonucleoprotein are <u>not</u> protective because it resides within the envelope and is not exposed to the blood. **Envelope** contains two types of glycoprotein spikes: **H** and **N**. One spike contains a hemagglutinin (H) and the other contains neuraminidase (N). A **matrix protein (M)** exists around the nucleo-capsid and helps support the envelope. It also has a role in viral budding.	1. Attachment via the hemagglutinin protein to cell surface receptors (sialic acid for A and B) allows entry into the cell. 2. Uncoating and translocation of the genome to the nucleus. 3. Translation of the (-) RNA to (+) stranded mRNA. The viral mRNA steals a 5' cap from host cell mRNA to initiate translation on cell ribosomes. 4. mRNA is translated by host cell ribo-somes into viral proteins. 5. Spike proteins are transported to the plasma membrane to become integral membrane proteins. 6. (+) RNA is also used as a template for **RNA dependent RNA polymerase** to create more (-) RNA for progeny viri-ons. This occurs in the nucleus. 7. The capsid proteins assemble around the segmented (-) RNA and RNA poly-merase. 8. Final assembly of the virion occurs when the envelope is acquired by bud-ding through the plasma membrane containing spike proteins.

Orthomyxovirus

RNA Viruses

Orthomyxovirus, Influenza

pathogen	structure, antigens and enzymes	diseases	notes

Orthomyxovirus
 	
Influenza A virus
 	
Influenza B virus
 	
Influenza C virus
	RNA dependent RNA polymerase is carried by the virus within the capsid.
	
	Spike proteins
	
	Hemagglutinin agglutinates RBCs. It binds sialic acid on host cells and is the first step in infection. Antibodies against the hemagglutinin spikes are protective. 12 antigenic types are known.
	
	Neuraminidase enzymatically breaks down neuraminic acid which is a component of mucus. This allows the virus to infect the upper and lower respiratory tract and is important in allowing new virions to leave the host cell. Antibodies against the neuraminidase do not prevent disease but reduce the severity. Not found on Influenza C. 9 antigenic types are known.
	
	The spike proteins, hemagglutinin and neuraminidase, are **type specific antigens** which have the ability to change their antigenic appearance. Worldwide epidemics occur when the antigenicity of the virus is changed enough to evade preexisting immunity.
	
	Antigenic shift is a <u>major change</u> which occurs <u>infrequently</u>. It is caused by the exchange of an entire segment of RNA with a different influenza virus (i.e. swine, horse or duck influenza A) that has infected the same cell. Influenza A is the only group which has antigenic shifts; these occur once per decade and cause a major epidemic.
	
	Antigenic drift is a <u>minor change</u> which occurs <u>frequently</u>. It is caused by random mutation of the RNA genome; does not cause epidemics.
		Influenza (the flu)
		   Transmission is by respiratory droplets or contact with secretions. The virus colonizes the respiratory tract but rarely becomes systemic.
		
		   Seasonal: infection usually occurs from December to March.
		
		   Incubation is 1-4 days.
		
		   Symptoms of infection include fever, headache, muscle pain, sore throat, dry cough and malaise.
		
		Pneumonia may occur as a progression of initial infection. Characterized by shortness of breath, hypoxia and diffuse infiltrates on x-ray.
		
		Secondary bacterial pneumonia is a common and serious consequence of influenza pneumonia. Influenza virus kills the ciliated cells of the respiratory tract which decreases host resistance to bacteria. Common secondary invaders are: S. *aureus*, S. *pneumonia* and H. *influenza*.
		
		Reye's Syndrome can be a late complication in children after influenza B or chicken pox.
		
		It is characterized by encephalopathy and hepatitis. Has been associated with aspirin use during the initial illness.
		
		Epidemics: Influenza A and B both cause epidemics but A causes more severe epidemics because of antigenic shifts.
		
		Influenza C does not cause epidemics. Infection with influenza C only causes minor respiratory symptoms.
		
		B and C only infect humans while A can infect humans, horses, pigs and birds.
			Treatment:
			Amantadine is an antiviral drug which is only effective against influenza A (types B and C are not affected).
			
			The drug prevents uncoating of the virus once it has entered the cell and is used to prevent and treat influenza infections.
			
			Prevention:
			Vaccine contains killed influenza A and B viruses. Yearly boosters are required to update the vaccine to the latest type specific antigens.
			
			The vaccine is recommended for persons at risk for respiratory infection, such as persons over the age of 65 or persons with chronic disease.
			
			Prophylaxis with amantadine can be used after exposure in certain situations (i.e. adjunct to vaccination for those at risk (immunocompromised, nursing home residents): if epidemic strain different from vaccine strain). Vaccine is the primary prevention.

Notes

RNA Viruses

Paramyxovirus

The Microbiology Companion, Topf and Faubel ©1997 **128**

Paramyxoviruses

Paramyxoviruses differ from the orthomyxoviruses by having <u>a single (-) RNA genome rather than a segmented one</u>. There are four medically important viruses in this family: measles virus, mumps virus, respiratory syncytial virus (RSV) and parainfluenza virus. The envelope contains various combinations of three protein spikes: neuraminidase, hemagglutinin and fusion protein.

pathogen	structure and enzymes	life cycle	Paramyxovirus	spike proteins
Paramyxovirus	SS (-) RNA not segmented helical nucleocapsid envelope RNA dependent RNA polymerase (L-protein)	1. Attachment via the hemagglutinin protein to cell surface receptors allows entry into the cell. 2. Uncoating and transcription of the (-) RNA to (+) RNA with a viral RNA dependent RNA polymerase occurs in the cytoplasm. The (-) RNA genome is transcribed into multiple mRNAs, one for each protein. The virus is able to cap its own mRNAs (unlike orthomyxovirus). 3. The mRNA is translated into viral proteins: spike proteins and RNA dependent RNA polymerase. 4. Spike proteins enter the RER and golgi membranes of the host. 5. A special (+) RNA strand is created as a template for RNA dependent RNA polymerase to create (-) RNA genomes for progeny virions. Normally (+) RNA contains code for only one protein (see step 2), this template (+) RNA is a complete genome. This occurs in the nucleus. 6. The capsid proteins assemble around the (-) RNA and RNA polymerase. 7. The envelope is acquired by budding through the plasma membrane which contains the spike proteins.	Measles Parainfluenza Mumps Respiratory syncytial virus Orthomyxovirus Influenza A, B, C	**H** and **F** **HN** and **F** **HN** and **F** **F** **H** and **N**

pathogen	structure/proteins	diseases	notes
Measles virus	SS (-) RNA not segmented helical nucleocapsid RNA dependent RNA polymerase Envelope contains 2 types of protein spikes: **H** and **F**. One spike contains hemagglutinin (H) and the other contains fusion protein (F). The fusion protein merges cells and is a hemolysin. Only one serotype exists.	**Measles** (rubeola, 7-day measles and hard measles) **Transmission** via respiratory droplets. Initially the virus infects the respiratory tract and produces symptoms of fever, runny nose and cough which occur after an incubation period of 1-2 weeks. Conjunctivitis and photophobia may also occur. **Koplik spots** appear in the mouth. They are bright red with a white spot in the middle and are diagnostic for measles. Unlike influenza virus, **measles virus enters the bloodstream and becomes systemic**. It infects the RES and causes vasculitis which contributes to the rash. The **rash** begins at the head and moves down. It contains **multinucleated giant cells** formed by the fusion protein. **Encephalitis** occurs in 1/1000 cases of measles. Mortality is 10% and permanent damage occurs in half. **Subacute sclerosing panencephalitis (SSPE)** occurs years after the primary infection and is uniformly fatal. It is due to a persistent infection by defective measles virus which cannot complete their life cycle. The disease is characterized by numerous inflammatory lesions in the brain. It is very rare.	Epidemics occur every two or three years. Recovery is associated with life long immunity. **Vaccine** contains **live attenuated virus** and is routinely administered to infants in conjunction with the vaccines for mumps and rubella (MMR). The vaccine has changed measles from a common childhood illness to a rare disease. In 1960 there were 500,000 cases, in 1988 there were 3500 cases. Recently the disease has had a resurgence. The vaccine does not always take with a single dose (2-5% failure) and many kids never receive follow up vaccinations. The disease now strikes college students and the immunocompromised.
Parainfluenza Four serotypes exist in humans and one in monkeys.	SS (-) RNA not segmented helical nucleocapsid Envelope contains 2 types of protein spikes: **HN** and **F**. One spike contains both hemagglutinin (H) and neuraminidase (N) and the other contains fusion protein (F).	**Croup** [type 1, 2] (laryngotracheobronchitis) is characterized by stridor (high pitched noisy respiration signaling respiratory obstruction) and a barking, harsh cough (croup). Rest, fluids and inhalation of steam assist recovery. Epidemics occur every other autumn. **Pneumonia** [type 3] is rare. **Upper respiratory infections** (i.e. pharyngitis and cold-like illness) occur in <u>adults</u>.	There is no vaccine. There is no specific antiviral therapy. Treatment for croup includes anti-inflammatory steroids and racemized epinephrine.

Paramyxovirus, Parainfluenza

RNA Viruses

RNA Viruses

pathogen	structure/proteins	diseases	notes
Mumps virus	SS (-) RNA not segmented helical nucleocapsid RNA dependent RNA polymerase Envelope contains 2 types of protein spikes: **HN** and **F**. One spike contains both hemagglutinin (H) and neuraminidase (N) and the other contains fusion protein (F). The fusion protein merges cells and is a hemolysin. One antigenic type.	**Mumps** **Transmission** is via respiratory droplets. **Incubation:** 18-21 days. The infection begins in the upper respiratory tract and then enters the blood to infect the parotid glands, testicles, ovaries, pancreas and/or meninges. **Painful swelling of the parotids** is the most characteristic symptom and occurs after a prodrome of fever, malaise and loss of appetite. Infection typically resolves in 10 to 15 days. **Orchitis** (infection of the testicles) is very painful in postpubertal males (the tunica albuginea does not stretch to allow swelling) and can result in <u>sterility</u> if infection is bilateral. Usually it is unilateral. **Meningitis** is self limiting and has no sequela. Mumps virus is one of the three most common causes of aseptic meningitis (culture negative CSF). The others causes are coxsackie and echo viruses.	Recovery is associated with life long immunity due to antibodies against the hemagglutinin spike. **Vaccine** contains **live attenuated virus** and is routinely administered to infants in conjunction with the vaccines for measles and rubella (MMR). A skin test is available for detection of past infection or vaccination. It is used most commonly to determine whether one's cell mediated immunity is intact. A negative skin test indicates a compromised cell mediated immune system.
Respiratory syncytial virus (RSV)	SS (-) RNA not segmented helical nucleocapsid Envelope contains one type of protein spike: **fusion protein (F)**. RNA dependent RNA polymerase Two antigenic types.	**Respiratory tract infections** **Transmission** is via respiratory droplets. **Lower respiratory tract infections** (i.e. pneumonia and bronchiolitis) occur in <u>infants</u>. These infections are widespread (1 in 250 infants is hospitalized with RSV) and may be fatal. **Upper respiratory tract** infections occur in children and adults. **Fusion protein** causes cells to fuse producing <u>multinucleated syncytia</u> which allow the virus to spread without exposing itself to antibodies.	Outbreaks of RSV occur every winter so that everyone has been infected by age 3. Nosocomial infections in nurseries are a big problem. **Wash your hands!** Recovery from infection does <u>not</u> induce immunity. Repeat infections are common. There is no vaccine. **Ribavirin**, an antiviral, is given via aerosol to infants on ventilators 2° to RSV.

Paramyxovirus, Respiratory Syncytial Virus

RNA Viruses

Togavirus family

Togaviruses have a (+) RNA genome which is an infectious genome because if it is introduced into the cell without any viral proteins the virus will be active. The RNA genome acts as its own mRNA. (-) RNA viruses need an RNA dependent RNA polymerase, a protein not found in host cells to produce mRNA for ribosomal translation. Togaviruses include four genera, two of which infect humans: Rubivirus (rubella virus) and Alphavirus (eastern equine encephalitis virus and western equine encephalitis virus). Alphaviruses are arboviruses (transmitted by insects).

Life cycle of a (+) RNA virus (infectious genome)

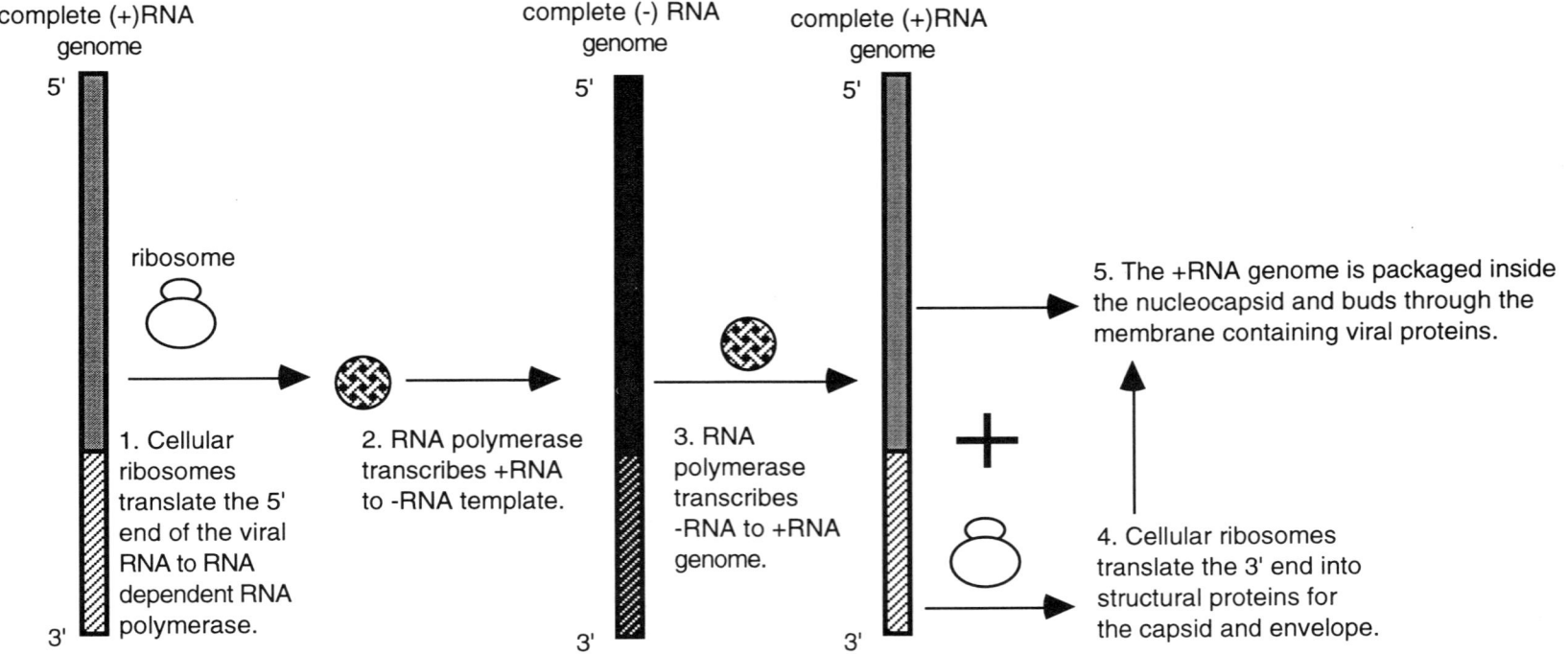

pathogen	structure and enzymes	diseases	life cycle
Rubella virus	SS (+) RNA, an **infectious genome** **Icosahedral** nucleocapsid is made of a single protein, the **C protein**. **Envelope** contains **hemagglutinin** protein spikes. Antibodies against hemagglutinin are protective. Only one antigenic type exists. Diagnosis is serologic. **IgM antibodies** indicate recent infection. If found in a pregnant woman she should undergo amniocentesis to check for virus in the amniotic fluid. **IgG antibodies** indicate protective immunity from either past infection or immunization.	**Rubella** (German measles) **Transmission** is by aerosol droplets. **Incubation**: 14-21 days. The initial infection is in the nasopharynx and cervical nodes. Organisms entering the blood stream deposit in tissues throughout the body and can cross the placenta. In adults, the infection causes a mild illness characterized by fever, malaise and a **maculopapular rash** which starts on the head and moves down the body. The rash remains for about three days. Transient arthritis is a frequent complication. Recovery induces life long immunity. **Congenital rubella syndrome** occurs when a pregnant woman is infected. Infection is most severe in the first trimester, and even worse in the first month. The syndrome can consist of: • heart: patent ductus arteriosus (PDA) • eyes: cataracts, glaucoma • brain: mental retardation, cerebral palsy, microcephaly, deafness • death of the fetus The child can shed the virus for up to 2 years after birth becoming a health threat to pregnant women and a reservoir of disease. 10-15% of women of childbearing age are susceptible to rubella.	1. Enters the cell by attaching via the hemagglutinin protein to cell surface receptors. 2. (+) RNA is translated by host ribosomes to produce viral proteins (i.e. spike proteins, RNA dependent RNA polymerase) 3. Spike proteins enter the RER and golgi to become integral membrane proteins. 4. (+) RNA is used by a viral encoded RNA dependent RNA polymerase to produce (-) RNA which is used as a template for the same RNA polymerase to make more (+) RNA genomes. This occurs in the cytoplasm. 5. The capsid proteins assemble around the genome. 6. The envelope is acquired by budding through the membrane where the spike proteins have been inserted. **Vaccine** contains live attenuated virus and is routinely administered to infants in conjunction with the measles and mumps vaccine. The vaccine should not be given to pregnant women or women who plan on getting pregnant in the next few months.

Togavirus, Rubella

The Microbiology Companion, Topf and Faubel ©1997 **133**

RNA Viruses

Togavirus, Alpha virus

The following togaviruses and flaviviruses are **arboviruses** (arthropod born viruses), which means they are transmitted by insects. The viruses can multiply in both human and insect hosts. In a human host, the virus causes cell death after release of viral progeny. In insects, the virus causes a latent infection in which the cells continue to shed viral progeny without dying. In most cases, humans are not reservoirs of disease because the virus never reaches high enough concentration in the blood to assure reliable transmission when the human host is bit by another insect. Two **exceptions** to this are **dengue fever virus and yellow fever virus**. Arboviral diseases only occurs in the summer months when insects are active.

Togaviridae, Alpha virus

pathogen	structure	diseases	vaccines and vectors
Eastern equine encephalitis virus	SS (+) RNA not segmented icosahedral nucleocapsid Envelope with protein spikes made of **hemagglutinin**. Antibodies against the hemagglutinin are protective.	**Eastern equine encephalitis (EEE)** is the most severe disease caused by the arboviruses. Humans and horses are dead end hosts. The disease is characterized by a sudden onset of headache, nausea, vomiting and fever. The fever often peaks twice: the first peak is due to general viremia (virus in the blood) and the second occurs when the virus invades the brain. Mental status changes (confusion) are quickly followed by seizures, coma and death. The fatality rate is 50%. Although recovery is associated with life long immunity, persons who recover commonly have life long neurologic damage.	There is no human vaccine available (a killed virus vaccine is used to protect horses). **Mosquitoes** are the vector and **wild birds** are the reservoir. Occurs in the eastern and southern regions of the U.S. The disease is rare with a handful of cases per year. Infection occurs during the summer, when mosquitoes are active.
Western equine encephalitis virus	SS (+) RNA not segmented icosahedral nucleocapsid Envelope with protein spikes made of **hemagglutinin**. Antibodies against the hemagglutinin are protective.	**Western equine encephalitis (WEE)** is similar to EEE, but less severe and causes fewer neurologic sequelae. Humans and horses are dead end hosts. Infection is common in endemic areas, but the illness is rarely symptomatic. Occurs in the western regions of the U.S. and in Canada. Occurs in the summer when mosquitoes are active.	There is no human vaccine available (a killed virus vaccine is used to protect horses). **Mosquitoes** are the vector and **wild birds** are the reservoir. The disease is more common than EEE and outbreaks occur with 100's of cases.

Flaviviridae family

Flaviviruses are smaller viruses than Togaviruses. They have SS (+) RNA with an icosahedral nucleocapsid made of **protein C**. The envelope contains protein spikes made of **protein E** and **protein M**. Antibodies against protein E are protective. All of the Flaviviruses have an arthropod vector. The Flaviviruses include: St. Louis encephalitis virus, yellow fever virus and the dengue fever virus.

pathogen	structure	diseases	vaccines and vectors
Saint Louis encephalitis virus	SS (+) RNA icosahedral nucleocapsid	**Saint Louis encephalitis** is the only arbovirus infection which occurs in urban areas. The disease is usually mild but outbreaks carry a 10% mortality rate and are more common in adults than children. The infection is widespread throughout the U.S. especially in Texas, Mississippi and Florida.	No vaccine or treatment is available. **Mosquitoes** are the vector and wild birds are the reservoir. Humans are dead-end hosts. Most common arbovirus encephalitis in U.S.
Yellow fever virus	SS (+) RNA icosahedral nucleocapsid one antigenic type	**Yellow fever** is an acute infection which causes liver damage and jaundice. Infection is characterized by the sudden onset of fever, chills, headache, nausea and vomiting. Infection occurs in central Africa, South and Central America and the Caribbean.	**Vaccine** contains live attenuated virus and is given to travelers and persons living in endemic areas. **Mosquitoes** are the vector and the reservoir is **monkeys** in the jungle and **humans** in cities.

Flaviviridae Family, Dengue fever virus

The Microbiology Companion, Topf and Faubel ©1997 **136**

pathogen	structure	diseases	vaccines and vectors
Dengue fever virus	SS (+) RNA icosahedral nucleocapsid four antigenic types all show cross reactivity	**Dengue fever** (break-bone fever) is characterized by headache, fever, backache, fatigue and severe pain in the joints and muscles. Two related complications are **dengue hemorrhagic fever (DHF)** and **dengue shock syndrome (DSS)**. DHF and DSS are characterized by inflammation, shock and hemorrhage. They occur when antibodies to one serotype of dengue fever virus (there are four) react with an infection caused by another serotype. The antibodies are unable to neutralize the infection and actually facilitate it by improving virus binding and entry into monocytes. Mortality is 10%. (see diagram)	**Mosquitoes** are the vector and the reservoir is **monkeys** in the jungle and **humans** in cities. Infection occurs in the Middle East, Africa and the Caribbean.

The infant has antibodies to one of the four serotypes of dengue fever virus. These can be maternal IgG or its own IgM.

The infant is then infected with a dengue fever virus of a different serotype.

Fc receptors

Monocyte

The antibodies partially cross react with the virus but are not effective enough to be protective. The antibodies bring the virus in contact with monocytes causing them to be preferentially infected. The monocytes release inflamatory mediators which can lead to hemorrhage, shock and life threatening complications.

Notes

RNA Viruses

pathogen	structure	diseases	notes
Hepatitis C virus (HCV) formerly grouped as non A-non B hepatitis	SS (-) linear RNA envelope This virus appears to be a member of the Flaviviridae family.	**Hepatitis** **Transmission** occurs through contact with infected blood, i.e. blood transfusions. Since the screening for HBV antibodies in donated blood has been implemented, HCV is the **most common cause of post-transfusion hepatitis**. **Incubation**: 2-26 weeks with a mean of 8 weeks. **Acute hepatitis** is clinically identical to the hepatitis caused by HBV and HAV. Symptoms include fever nausea, vomiting, jaundice and fatigue. **Chronic hepatitis** occurs in at least 50% of persons infected with hepatitis C. Chronic infection is generally less severe than in patients with chronic HBV. (please see page 92 for a comparison of Hepatitis C with the other viral causes of hepatitis)	Antibodies to the viral component **C-100-3 antigen** are present 15 weeks after infection. This antibody is evidence of **previous** or **chronic infection.** Alpha interferon is helpful in some patients with chronic infections. No vaccine; immunoglobulin therapy not effective. Blood for transfusion is tested for anti-HCV; which has reduced the risk of hepatitis C infection to about 0.6% per transfusion.
Hepatitis E virus (HEV) formerly grouped as non A-non B hepatitis	small SS RNA virus non-enveloped not yet classified cannot be cultured	Causes water borne **epidemics** of acute hepatitis. **Transmission** is fecal-oral or by contaminated water. **Incubation**: 3-7 weeks. **Acute hepatitis** is clinically identical to the hepatitis caused by HBV and HAV. Symptoms include fever nausea, vomiting, jaundice, and fatigue. No chronic infections. More severe in pregnant women, 20% mortality.	Epidemics are widespread in India, southeast Asia, Africa and Mexico.

Rhabdovirus

◆ Single Stranded (-) RNA
◆ bullet shaped capsid
◆ envelope

pathogen	structure	diseases	notes
Rabies virus a rhabdovirus	SS (-) RNA bullet-shaped nucleocapsid envelope RNA dependent RNA polymerase single antigenic type Infected neurons contain pink (eosinophilic) inclusions called **Negri bodies**. Immunofluorescence can detect the virus in brain tissue and corneal scrapings. This is used to determine if an animal which has bitten a person is infected.	**Rabies** virus can infect most mammals. **Transmission** is by the bite of an infected animal. Animals are infectious only the week before they die. Bats are unique in their ability to transmit the disease while remaining healthy. Rabbits and rodents are the only mammals unable to transmit the disease. The virus first multiplies at the site of the bite and then moves up peripheral nerves towards the CNS. It replicates in the CNS, causes encephalitis and travels back down peripheral nerves to the salivary glands. The incubation period varies (from two weeks to nine months) depending on how large the infectious dose is and how far the virus must travel to get into the CNS. Initial symptoms are nausea, vomiting, anxiety and pain at the site of the bite. These are followed by **confusion, lethargy and hydrophobia**. End stage disease is characterized by seizures, coma and death. **Hydrophobia** is the failure to drink because it is too painful. The failure to swallow secondary to pain leads to the classic characterization of rabies: " . . . foaming at the mouth."	Prevention and treatment in humans is of two types: Pre-exposure: **rabies vaccine** (human diploid cell vaccine, **HDCV**) is given to people at high risk of exposure. It contains an inactivated (killed) virus which has been grown in human diploid cells. Post-exposure: HDCV and **rabies immune globulin (RIG,** IgG) are administered at different locations on the body. The patient follows up with four additional doses. Cleaning of the wound as soon as possible is also very important. Live attenuated vaccine is used to immunize cats and dogs.

RNA Viruses

Bunyavirus

- ◆ three segments
- ◆ Single Stranded (-) RNA
- ◆ helical capsid
- ◆ envelope

pathogen	structure	diseases
Hantaviruses Hantaan Puumala Seoul virus Four Corners/Sin Nombre Virus Prospect Hill	SS (-) RNA three RNA segments envelope four polypeptides replication in cytoplasm envelope from Golgi	**Hemorrhagic fever with renal syndrome** is an acute illness which is characterized by nephritis (which may lead to acute renal failure), hemorrhage and shock. The disease is occurs in five stages: fever, hypotension, oliguria, diuretic and convalescent. Transmission is through contact with infected rodents. **Hantavirus pulmonary syndrome** is characterized by prodrome of fever, cough, myalgias (muscle aches). As the disease progresses, severe shortness of breath leading to respiratory failure occurs from pulmonary edema. A low platelet count also occurs. An outbreak of this syndrome occurred in 1993 in southwestern US. Of those affected, 60% died. Transmission occurs by inhalation of dust contaminated with the excrement of infected rats. No specific treatment exists other than supportive.

Filovirus

- non-segmented
- Single Stranded (-) RNA
- helical capsid
- envelope

pathogen	structure	disease	notes
Ebola virus Marburg virus	SS (-) RNA envelope long, filamentous morphology five polypeptides replication in cytoplasm envelope from cell membrane	**Hemorrhagic fever** is characterized by the acute onset of high fever associated with hemorrhage throughout the body. Bleeding occurs into the skin and from the nose, GI and genitourinary tract. Low platelet count also occurs. The diffuse hemorrhage often leads to shock and death. The disease is endemic in Africa. Transmission occurs through contact with an infected animal (monkeys);. Person to person spread occurs through contact with infected bodily secretions.	Ebola virus is one of the most lethal human pathogens known. Mortality from infection is greater than 70%.

RNA Viruses

RNA Viruses

Retroviruses

- diploid SS RNA
- icosahedral capsid
- envelope

Retroviruses can cause cancer or immune deficiency in humans. They are spherical, enveloped RNA viruses and carry a viral encoded RNA dependent DNA polymerase known as **reverse transcriptase**. The virus contains two copies of its genome so it is a **diploid virus**. The virus contains unique tRNAs which serve as primers for reverse transcriptase. The life cycle of the retroviruses is very complex:

1. **Fusion** of viral envelope with host-cell membrane. In HIV the gp120 protein binds CD4$^+$ T-cells.

2. Viral capsid enters the host cell and **uncoats**. The virus releases both copies of the RNA genome and the GAG and POL proteins.

3. **Viral RNA is transcribed into DNA** by RNA dependent DNA polymerase (reverse transcriptase).

4. Reverse transcriptase produces DS DNA with long terminal repeats (LTRs) at either end. Reverse transcriptase destroys the RNA template while producing the DNA.

5. The DS DNA enters the nucleus and is **integrated into the host cell genome**. This requires the viral protein integrase which is coded by the *pol* gene. Viral DNA integrated into the host cell genome is called a provirus.

    Sometimes the DNA remains free in the nucleus this is called an episome.

6. The viral genome can remain dormant within the host indefinitely.

7. The viral **DNA is transcribed to mRNA** by the normal host RNA polymerase under the influence of a promoter and enhancer contained in the viral DNA.

8. This mRNA is then spliced by post-transcriptional processing. If the mRNA is spliced into small pieces regulatory proteins will be produced. If the same mRNA is spliced into long segments then structural proteins will be produced.

9. The **mRNA is translated** by ribosomes into viral proteins.

10. Two membrane proteins, coded by the viral gene *env*, travel through the RER and golgi to insert into the cell membrane.

11. Capsid proteins, coded by the viral gene *gag*, are produced by cytoplasmic polyribosomes and self-assemble around the viral genome, reverse transcriptase and integrase.

12. The **virus buds through the membrane** containing the viral membrane and does not lyse the cell.

RNA Viruses

The retroviral genes can be divided in to two categories: structural and regulatory.

structural gene	structural protein	function
gag Group *antigen* codes a primary polypeptide, Pr*gag,* which is processed into the internal core proteins.	matrix (MA) capsid (CA) or **p7** core (NC) or **p24**	Found between viral capsid and lipid envelope. p24 is genetically stabile, antibodies to p24 are used to document infection but do not neutralize the virus.
pol	reverse transcriptase (RT) protease (PR) integrase (IN or Int)	Reverse transcriptase is multi functional: • Translates viral RNA to DS DNA for insertion into the host genome. • RNAse activity which destroys the original RNA template while producing DNA strand. This prevents RNA-DNA hybrids. • The HIV reverse transcriptase is error prone which accounts for the *high mutation rate* of this virus. Protease acts to cleave functional proteins from protein precursors produced by cellular ribosomes. Targeted by the newest anti-retroviral drugs. Splices the viral DS DNA into the host genome.
env *env*elope genes are very heterogeneous. They mutate easily giving the virus its chameleon like character.	surface (SU, gp 120) transmembrane (TM, gp41) *env* actually codes for a precursor protein, gp160, which is cleaved into gp120 and gp41	Allows the cell to bind to host cells. Attaches to cells containing the surface molecule CD4 (helper T-cells among others). Antibodies to gp-120 neutralize the virus but the protein mutates rapidly making this a poor way to fight the disease. Aids viral fusion with host cell and is important in infecting cells without CD4 such as glial cells and fibroblasts. Also forms syncytia allowing transfer of the virus without exposing it to antibodies.

regulatory gene	regulatory protein	function
tat HIV *tax* HTLV	TAT TAX transactivating factor	**Strong transactivator**. These *antilatency genes* encode proteins found in the nucleus of the host cell and act as activators of viral DNA transcription. They can also increase expression of cellular genes, including oncogenes causing cancer. Decreases the expression of class I MHC proteins which reduces the ability of cytotoxic T-cells to kill HIV infected cells. Cytotoxic T-cells normally recognize viral antigens in association with class I MHC proteins and then kill the infected cell preventing progression of the virus.
rev HIV *rex* HTLV	REV REX regulator of virion protein	These genes codes for proteins which promote the translation of viral mRNAs other than *rex* /*rev* and *tax* /*tat*. They control latent infections. The structural proteins are coded for by very long mRNAs which have difficulty leaving the nucleus. REX/REV facilitate the transport of the intact mRNAs out of the nucleus. When the long mRNAs are sliced into shorter strands, they code for regulatory proteins.
nef	NEF negative regulatory factor	Initially thought to down regulate transcription. Recent in vivo evidence shows increased viral transcription. It also causes the cell to decrease the expression of CD4 on the cell surface preventing reinfection of an already infected cell. NEF acts near the promoter of the viral DNA to inhibit transcription.
vif	VIF viral infectivity factor	Antilatency gene. Helps viral maturation and budding.
vpr	viral protein R	involved in the infection of monocytes and macrophages
vpu	viral protein U	involved in the infection of monocytes and macrophages
LTR	does not code for a protein but is essential to initiate protein synthesis	**transcription initiating segment** the 5' end has a binding site for the TAT protein which enhances viral mRNA transcription.

RNA Viruses

Human Immunodeficiency Virus

pathogen	structure / enzymes	pathophysiology	diseases
Human immuno-deficiency virus HIV **HIV-2** is a retrovirus which causes AIDS. Only a very few cases have been found but it is increasing in prevalence in Africa. It causes a less severe form of AIDS than HIV-1.	retrovirus diploid SS RNA icosahedral capsid lipid envelope contains two viral specific glycoproteins **gp120** and **gp41** reverse transcriptase Screening tests use **ELISA**, which detects antibodies to HIV. Because of the possibility of false positives with ELISA, the **Western immunoblot** is used for confirmation. The Western blot detects antibodies to HIV polypeptides. Please read <u>And the Band Played On</u> by Randy Shilts. It is an excellent history of the AIDS epidemic which carefully examines the political, scientific and social issues which were largely ignored by the popular press.	The HIV envelope protein gp 120 binds to the surface molecule CD4 on host cells. Any cell with CD4 (macrophages helper T-cells, and monocytes) can be infected by HIV. After infection the viral RNA is converted to DNA and integrates into the host cell genome and becomes latent (no active replication viral replication). Note: latency is used to describe two different events in HIV/AIDS: 　Latency described above refers to the activity of a single virally infected cell which will not be actively producing virions. 　Clinical latency describes the time after acute viral infection and before the onset of AIDS. During this period their is active viral reproduction but it is mostly limited to the lymph tissues and there is little evidence of viral production in the peripheral blood . The virus has a long latency with an average time of 11 years between infection and death. Some individuals have been HIV + for 14 years and are still alive. At some time the virus becomes activated and the cell begins to produce virions which bud through the cell without causing cell lysis. After the virions are released the cell dies by an unknown mechanism. The virus may activate programed cell death: **apoptosis**. The virus can cause the production of giant syncytia made of many CD4 cells. This also results in cell death.	The helper T cells are involved in both humoral and cellular immunity so the selective destruction of this cell line leads to a combined immunodeficiency which results in a wide variety of opportunistic infections. In addition to immunodeficiency HIV has direct neural toxicity. Can cause a polyneuropathy, myelopathy and a encephalopathy causing dementia. The CNS disease results from infection of microglial cells, monocytes and macrocytes

Disease	Transmission	diagnosis	natural history

Acquired Immunodeficiency Syndrome (AIDS)	**Transmission** occurs through the exchange of body fluids.
	
	CDC HIV/AIDS Surveillance Report
	
	♂♂ sexual contact — 50%
	IV drug use (IVDU) — 26%
	♂♂ sex and IVDU — 6%
	heterosexual contact — 9%
	hemophilia — 1%
	blood transfusion — 1%
	Other/not reported — 7%
	total cases: — 573,800
	
	data is cumulative *total* AIDS cases in the US for adults/adolescents.
	
	Children: 90% mother with AIDS or HIV risk factor. total: 7,629 cases
	
	Globally, the vast majority of new cases are due to heterosexual intercourse.
	
	transmission ♂ ➜ ♀ is 20x more efficient than ♀➜♂
	
	Healthcare workers are at increased risk but still only 1 in 400 needle sticks with HIV+ patients result in seroconversion. Ziduvatine after needle sticks is being used to decrease the chance of infection. The efficacy of this is unproven.
	
	leading cause of death for men 25-44 since 1992.
	
	4th leading cause of death among women since 1993.
		Acquired immunodeficiency syndrome (AIDS) is the major disease caused by the HIV. The syndrome is characterized by immunodeficiency, evidence of HIV infection and opportunistic infection or malignancy. AIDS defining illnesses are listed below:
		• Candida
		    bronchi, trachea or lungs or esophagus
		• Cervical Carcinoma, invasive
		• Coccidioidomycosis
		    extrapulmonary or disseminated
		• Cryptococcus, extrapulmonary
		• Cryptosporidiosis, extrapulmonary
		• CMV
		• Herpes simplex virus
		    esophagitis, pneumonia, chronic ulcerative
		• Histoplasma
		    disseminated or extrapulmonary
		• HIV dementia
		• HIV wasting
		• Isosporiasis, chronic interstitial
		• Kaposi's sarcoma
		• Lymphoma: brain, Burkitt's or
		    immunoblastic
		• Mycobacterium avium or kansasii
		• Mycobacterium Tuberculosis
		• Other mycobacterial disease
		• Pneumocystis carinii pneumonia
		• Pneumonia, recurrent
		• Progressive multifocal
		    leukoencephalopathy
		• Salmonella septicemia, recurrent
		• toxoplasmosis of the CNS
		• **Immunodepression: CD4 < 200**
		
		AIDS is diagnosis secondary to immunosuppression in 57% of cases and due to opportunistic infections in the remaining 43% of cases.
			HIV infection is classified into three clinical groups (A,B,C) and three CD4 count categories (1,2,3):
			
			Clinical categories:
			
			A Acute infection. The acute illness is mononucleosis-like and occurs weeks after infection in up to 50% of patients.
			
			Symptoms include lymphadenopathy, fever, rash, sore throat and myalgia.
			
			With resolution patients typically have persistent lymphadenopathy which are sites of rapid viral multiplication
			
			B Symptomatic patients but without AIDS defining illnesses. This could be oral/vaginal candidiasis, recurrent zoster, thrombocytopenic purpura.
			
			C AIDS by either defining illness or immune suppression.
			
			CD4 count categories:
			
			1 CD4 > 500 cells/mm^3
			
			2 CD4 200-499 cells/mm^3
			
			3 CD4 < 200 cells/mm^3

Human Immunodeficiency Virus

RNA Viruses

Human Immunodeficiency Virus   The Microbiology Companion, Topf and Faubel ©1997   **148**

Disease	Transmission	diagnosis	natural history
		Pneumocystis carinii pneumonia is by far the most common opportunistic infection at diagnosis, 39% of patients diagnosed with AIDS have PCP. **Kaposi's Sarcoma** (found mostly in homosexual men) is a neoplasm of blood vessels that causes mucocutaneous lesions. Thought to be caused by HSV 8 **AIDS dementia complex** is a degenerative neurologic disease with symptoms similar to Alzheimer's.	

The time and resources which have been poured into HIV research has begun to yield remarkable advances in the treatment of AIDS. The treatment of HIV/AIDS is focused in two major arms: anti-retroviral and anti-opportunistic infection.

treatment	What it is	significance	notes
viral load measurement	A lab test which allows measurement of the number of virions in the blood.	A major goal of modern HIV/AIDS treatment is maintaining a low viral load. This helps in two ways: • minimizes damage to the immune system so the patient can avoid opportunistic infections • a smaller viral population results in less genetic diversity in the virus. This minimizes the chance drug resistance will develop	The lowest viral load now measurable is 500 copies per ml. This is the *only* case in infectious diseases where the actual infectious agent is measured and quantified.
Combination therapy	the use of multiple anti-retro-viral drugs simultaneously	One of the major break through in treatment. Using multiple drugs with differing mechanisms of action has minimized the emergence of resistance and maximized the log kill.	
Log Kill	A method of measuring effectiveness of treatment by comparing viral load before and after treatment.	A "one" log kill decreases the viral load by one order of magnitude or one log: 10,000 to 1000 A log kill of "2" would drop a viral load from 10,000 to undetectable levels.	Most treatment regiments aim for a log kill of two or a drop to unmeasurable levels.
Reverse Transcriptase inhibitor	anti-viral drug which blocks the creation of DNA from the viral RNA template, one of the first steps in viral replication.	most of these are nucleosides which appear to the reverse transcriptase enzyme to be one of the four *correct* nucleosides (adenine, guanine, cytosine, thymine) but actually disrupt the enzyme. nevirapine is a non-nucleoside reverse transcriptase inhibitor.	Didanosine (ddI) Lamivudine (3TC) Stavudine (d4T) Zalcitabine (ddC) Zidovudine (ZDV) Nevirapine
Protease inhibitor	anti-viral drug which prevents the primary amino acid chains from being chopped into multi-ple mature viral proteins.	New class of anti-HIV drugs which have been very effective in lowering the viral load when used in combination with other anti-retroviral drugs.	Indinavir Nelfinavir Ritonavir Saquinavir
Ziduvudine (ZDV) formerly azidothymidine (AZT)	reverse transcriptase inhibitor severe side effects which include anemia and granulocytopenia (↓ white blood cell count)	It has been used with varying success in slowing the progression of the disease Currently considered first line therapy usually in combination with 3TC or ddI.	Decreases maternal-fetal transmission from 24% to 8%.

Human Immunodeficiency Virus

The prevention of opportunistic infections has extended the life of patients with AIDS:

Disease	prophylaxis	when to start	notes
Mycobacterium tuberculosis	isoniazid (INH) and pyridoxine Before starting prophylaxis, patients should be evaluated for active TB by CXR. Active TB needs at least a 3 drug regiment.	evaluate at diagnosis and prophylax if: • PPD > 5 mm • contact with active TB	rifampin is used for INH resistant *M. tuberculosis*. BCG is contraindicated because it can cause disseminated disease
Streptococcus pneumoniae	standard pneumococcal vaccine The vaccine is most effective if the patient has a CD4+ > 200/μL.	Should be given at the time of diagnosis.	Vaccination is associated with a transient increase in HIV activity of unknown significance
HPV and invasive cervical cancer	Two pap smears in the first year of diagnosis of HIV infections and then yearly after that.	A pelvic exam should be done at diagnosis and 6 months later and then yearly.	Abnormal pap smears need to be followed with repeat exam, colposcopy or directed biopsy
Hepatitis B virus	Hepatitis vaccine	The vaccine should be initiated at diagnosis.	
Pneumocystis carinii pneumonia	Trimethoprim-sulfamethoxazole (TMP-SMZ). Both double strength and single strength doses are effective.	**CD4+ < 200/μL**	TMP-SMZ also offers some protection against *H. flu* and *S. pneumo* and toxoplasmosis.
Toxoplasma gondii	Trimethoprim-sulfamethoxazole (TMP-SMZ). Use double strength tab once a day. Test for toxoplasma IgG at time of diagnosis and when the CD4+ < 100/μL	**CD4+ < 100/μL** only needed if patient has toxoplasma antibodies (IgG)	Seroconversion can be decreased by avoiding undercooked meat and stray cats. HIV patients should avoid contact with cat feces.
Histoplasma capsulatum	Itraconazole	**CD4+ < 100/μL** only recommended for areas endemic for histoplasmosis.	Itraconazole is a teratogen in animals so it should not be used in pregnant women.
CMV	oral gancyclovir has been used but severe side effects (neutropenia, anemia) and questionable efficacy make the choice to prophylax unclear	**CD4+ < 50/μL** only needed if patient has antibodies (IgG) to CMV	To prevent serious infection patient must know early signs of CMV retinitis: seeing "floaters" and decreased visual acuity.
Mycobacterium avium complex	azithromycin 1200 mg once a week clarithromycin 500 mg twice a day	**CD4+ < 50/μL**	

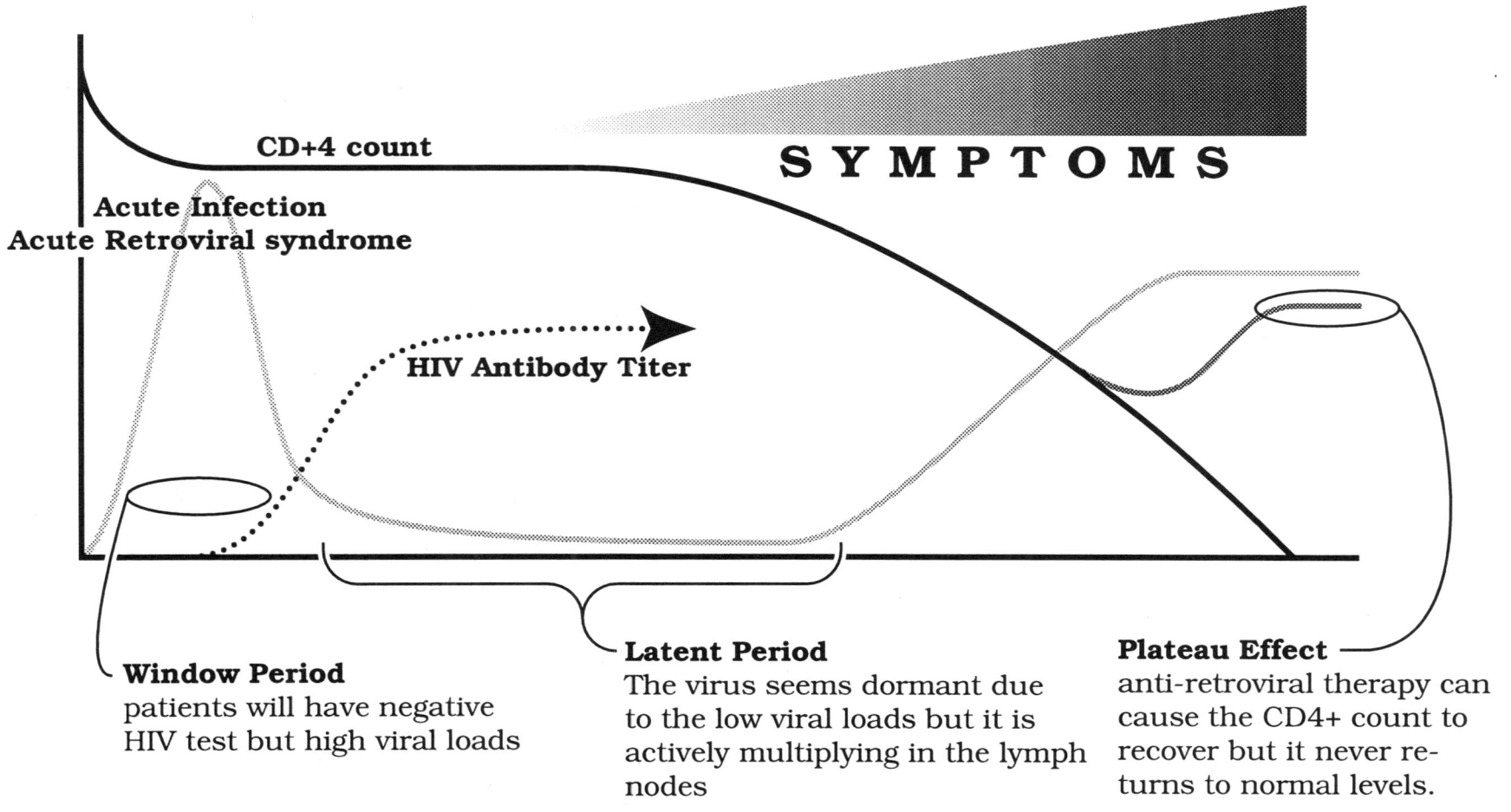

CD+4 count

Acute Infection
Acute Retroviral syndrome

S Y M P T O M S

HIV Antibody Titer

Window Period
patients will have negative
HIV test but high viral loads

Latent Period
The virus seems dormant due
to the low viral loads but it is
actively multiplying in the lymph
nodes

Plateau Effect
anti-retroviral therapy can
cause the CD4+ count to
recover but it never re-
turns to normal levels.

Human Immunodeficiency Virus

RNA Viruses

pathogen	structure, antigens and enzymes	diseases	notes
Human T-cell leukemia virus HTLV	retrovirus diploid SS RNA icosahedral envelope reverse transcriptase	**Leukemias and lymphomas** **T-cell malignancies** have been associated with HTLV primarily in Japan, the Caribbean and Africa. **Tropical spastic paralysis** is characterized by progressive weakness of the legs and lower body. Possible association with **multiple sclerosis**. Transformation results from the continual expression of the viral *tax* gene. Apparently, the TAX protein not only activates viral transcription, but also activates the expression of one or more cellular genes (possibly protooncogenes) which results in malignant transformation	**HTLV-II** has been isolated from patients with hairy-cell leukemia and is a retrovirus similar to HTLV-I HTLV was the first retrovirus ever discovered that infected people. The virus was discovered by Dr. Gallo the famous retrovirologist at the NCI.

Notes

RNA Viruses

Unenveloped RNA viruses are divided into two families: Picornaviruses and

Reoviruses. Because these families of viruses do not have an envelope, they are not inactivated by solvents such as ether, 70% alcohol or lysol.

Picornaviruses
◆ SS (+) RNA, not segmented.
◆ RNA contains a protein on the 5' end (VPg) which acts as a primer during RNA synthesis.
◆ Genome codes for 4 structural proteins (VP1-4) and the RNA associated VPg.
◆ Icosahedral nucleocapsids.
◆ Very small virus 22-30 nm (nanometers 10^{-9} meters).

Divided into two categories:

Enterovirus	**replicate at 37°C, stable at pH 3**
Poliovirus	poliomyelitis
Coxsackievirus	aseptic meningitis, upper respiratory infections
Echoviruses	aseptic meningitis
Hepatitis A	acute viral hepatitis
Rhinovirus	**replicate at 33°C, inactivated at pH 3**
over 100 serotypes	common cold

Reoviruses
◆ **double stranded** RNA, segmented
◆ doubled layered nucleocapsid

Rotavirus	gastroenteritis
Orbivirus	Colorado tick fever

	structure/enzyme	diseases	life cycle
Poliovirus	SS (+) RNA with 5' protein primer icosahedral nucleocapsid **Three serologic types** (1, 2, 3) are based on antigenic differences in the nucleocapsid. Protection from polio requires antibodies to all three types. no envelope grows at 37°C stable at low pH The virus can only infect primates.	**Poliomyelitis** **Transmission** is by the fecal-oral route. Infection begins in pharynx and GI. Spread throughout the body occurs via the blood and retrograde travel within neurons. The virus has a predilection to replicate within <u>anterior horn motor neurons</u>; **paralysis** occurs when these cells are destroyed. The majority of infections are asymptomatic, especially in infants. Older hosts have higher rates of paralysis. Symptomatic infections can be of 3 types: **Abortive polio myelitis** is a mild illness characterized by fever, headache, vomiting and spontaneous recovery. **Nonparalytic polio myelitis** appears as aseptic meningitis with fever, headache, vomiting and a stiff neck. **Paralytic polio myelitis** is the most severe form and is characterized by asymmetric flaccid paralysis with no sensory deficits. The paralysis worsens for a few days and then temporarily damaged neurons begin to recover. After 6 months, paralysis is permanent. Bulbar polio is a form of paralytic polio myelitis which results from brain stem involvement resulting in paralysis of the cranial nerves and the muscles of respiration.	The life cycle occurs entirely within the cytoplasm: 1. After binding to cell surface receptors (found only in primates) the virus enters and uncoats. 2. (+) RNA is translated into a single long protein called **noncapsid viral protein 00**. 3. Noncapsid viral protein is then cleaved to form the capsid proteins and the RNA dependent RNA polymerase. The polymerase first makes a (-) RNA template before transcribing the (+) RNA genome. 4. The capsid assembles and is released by cell lysis. (No envelope, no budding.)
The Polio Vaccines	Neither Salk or Sabin received the Nobel prize for their contributions to world health.	Two **vaccines** are available. The Salk has recently been recommended in the US due to the low prevalence of the wild type polio. The only indigenously acquired cases of polio in the last 17 years were due to revision of the attenuated virus. The risk was estimated at one case of paralytic polio for every 2.4 million doses. <table><tr><td>**Salk Vaccine** • killed virus • all three serotypes • given by injection • induces IgG in the blood • does not induce secretory IgA • no possibility of reversion • does not need refrigeration</td><td>**Sabin Vaccine (TOPV)** • live attenuated virus • all three serotypes • given orally • induces IgG and secretory IgA • replicates in the GI and is shed in the feces providing community inoculation • reversion is possible and causes vaccine associated paralytic polio (VAPP) • must be kept refrigerated</td></tr></table>	

RNA Viruses

RNA Viruses

pathogen	structure and enzymes	diseases	notes
Coxsackievirus	SS (+) RNA with 5' protein primer icosahedral nucleocapsid no envelope grows at 37°C stable at low pH Divided into group A and B based on their pathogenicity in mice: **Coxsackie A**: widespread, inflammatory reaction and necrosis of skeletal muscle, leading to flaccid paralysis and death. **Coxsackie B**: encephalitis, resulting in spasticity and convulsions. Life cycle is similar to poliovirus.	**Aseptic meningitis** may be caused by either type A or type B. Lasts 5-14 days and is normally mild. Coxsackie is one of the leading causes of aseptic meningitis (as are echo and mumps virus). **Transmission** is by the fecal-oral route and respiratory droplets. Occurs most frequently in the summer and fall. **Group A viruses** attack skin and mucous membranes. **Herpangina**: fever and sore throat. **Hand-foot-and-mouth disease**: vesicular rash on the hands and feet and ulcerations in the mouth. Occurs in children. **Group B viruses** attack organs, i.e. heart and liver. **Pleurodynia**: fever with severe pleuritic pain (pain during inspiration). **Myocarditis**: fever, chest pain and the signs of congestive heart failure (peripheral edema, dyspnea, exercise intolerance). Infection in pregnant women can cause congenital heart defects in the fetus. Can be fatal in neonates. Possible role in the pathogenesis of **Type I** (juvenile onset) **diabetes** following viral pancreatitis.	no vaccine Most common cause of non-bacterial CNS infections. 23 serotypes of coxsackie A. 6 serotypes of coxsackie B.
Echovirus	SS (+) RNA with 5' protein primer icosahedral nucleocapsid no envelope grows at 37°C stable at low pH 30 serotypes have been isolated life cycle is similar to poliovirus	**Aseptic meningitis** Transmission is by the fecal-oral route. Echovirus is one of the leading causes of aseptic meningitis (as are coxsackie and mumps). Infantile diarrhea. Maculopapular rash with petechia.	ECHO stands for enteric cytopathic human orphan. An orphan virus is not associated with any disease. The name is a misnomer because echoviruses do cause disease in humans. No vaccine available.

RNA Viruses

pathogen	structure and enzymes	diseases	notes
Hepatitis A Virus (HAV) also known as: enterovirus 72	SS (+) RNA with 5' protein primer icosahedral nucleocapsid no envelope grows at 37°C stable at low pH only one serotype life cycle similar to poliovirus Diagnosis is by detecting IgM antibodies to HAV.	**Acute hepatitis** occurs after a short incubation period (20- 25 days) compared to other hepatitis viruses. Transmission is by the **fecal-oral** route. **Incubation**: 14-40 days. The virus multiplies in the intestinal mucosa and is shed for up to two weeks before symptoms appear. Spread to the liver is via the blood stream. **Symptoms** of infection include fever, loss of appetite, nausea, vomiting and jaundice. Mild undiagnosed cases are common (anicteric hepatitis A) and age related. Younger patients have fewer symptoms. **Diagnosis** is by the detection of anti-hepatitis A IgM antibody. Always resolves spontaneously with <u>no chronic infection</u> or carrier state. No specific treatment other than supportive.	30% of American adults have been infected with HAV. Recovery provides life-long immunity. **Prevention**: **Vaccine** contains inactivated hepatis A virus. Can be given to persons travelling to endemic areas. Official recommendations for use are currently not available. **Immune serum globulin** (ISG) is used for immediate protection after exposure.
Rhinoviruses	SS (+) RNA with 5' protein primer icosahedral nucleocapsid no envelope **over 100 serotypes** Grows at 33°C, the temperature of the nasal mucosa. Destroyed by stomach acid. Life cycle similar to poliovirus.	**The common cold** is an upper respiratory infection characterized by a runny nose, sneezing, sore throat, cough and/or headache. Symptoms usually last about one week. **Transmission** is by contact with nasal discharge (snot) on fomites followed by hand-nose contact. Rhinovirus is the primary cause of the common cold which occurs most often during spring and fall. Due to temperature growth restriction, these infections cannot become systemic.	Children get more colds then adults because infection causes type specific immunity via IgA. There is no vaccine and treatment is symptomatic. To prevent infection: wash your hands and don't touch your nose.

Notes

RNA Viruses

Reoviruses

- ◆ segmented **DS** RNA
- ◆ double layered icosahedral capsid
- ◆ no envelope

pathogen	structure and enzymes	diseases	notes
Rotavirus member of the Reovirus family	segmented DS RNA There are 11 segments of RNA which each code for a single mRNA. Double layered icosahedral nucleocapsid which looks like a wheel. no envelope four serotypes RNA dependent RNA polymerase	**Gastroenteritis** **Transmission** is via the fecal-oral route and possibly by contaminated water and food(?). The virus replicates and is cytopathic in the intestinal mucosa. The virus does not invade. Infection is characterized by nausea, **vomiting**, and watery, non-bloody diarrhea. Rotavirus is the primary cause of gastroenteritis in in-fants (50% of hospitalized children 6-24 months have Rota), in whom the disease can be very severe. Infection has a high mortality rate in developing countries.	No vaccine exists. Treatment is symptomatic.
Orbivirus member of the Reovirus family (is also an arbovirus: transmission is by insects)	segmented DS RNA There are 10 segments of RNA which each code for a single mRNA. Double layered icosahedral nucleocapsid which looks like a wheel. no envelope RNA dependent RNA polymerase	**Colorado Tick Fever** (CTF) **Transmission** is by the wood tick which is carried by the small mammals of the Rocky Mountains. Symptoms of infection include fever, headache and muscle pain. May be able to cross the placenta and cause congenital defects or fetal death.	No vaccine exists. Treatment is symptomatic.

Think: **Reo**viruses are the only RNA viruses with double stranded RNA (**re**peat**o** viruses).

Prions: proteinaceous infectious particles

Prions differ from all other infectious organisms by their lack of *any inherent genetic material.*
Prions are a newly discovered and controversial form of infectious and genetic disease. These particles are proposed to be the cause of the spongiform encephalopathies. The prions are slightly proteins that are nearly identical in amino acid composition to their natural counterparts produced by the cell but they are folded in a totally different confirmation. One of the critical attributes of the prion is the ability to induce a conformational change in normal cellular protein to the altered and pathologic protein.

pathogen	structure	life cycle	Pathology

Prion	**PrPc**, the normal cellular protein is folded into a secondary structure which has four subunits made of alpha helixes. The gene for PrPc is on chromosome 20.
	
	PrPsc, the scrapie protein or the actual infectious particle is folded into a secondary structure of ß-pleated sheets.
	
	PrPsc resists break down by normal cellular proteases while PrPc is broken down like any normal protein by proteases.
	
	PrPsc in Gerstmann Strauessler-Scheinker disease differs from PrP by a single amino acid: proline → leucine.
	
	When PrPc and PrPsc are mixed in a test tube the PrPsc converts the normal cellular PrPc to disease causing PrPsc.
		1. Inoculation of the prion protein (PrPsc) can be by oral, intravenous or most effectively by intrathecal contact
		  A. This can be from ingestion of contaminated meat
		  B. iatrogenically through contaminated surgical instruments or transplants:
		    a. corneal
		    b. dura mata
		    c. human growth hormone derived from cadaveric pituitary glands
		  C. cannibalism of infected brains
		  D. endogenous production of the PrPsc protein (autosomal dominant form of disease)
		2. The PrPsc protein enters the neurons of the CNS
		3. Induces healthy endogenously produced cellular PrP (PrPc) to change conformational shape.
		4. The new conformed PrPsc also begins converting the PrPc to PrPsc
		5. PrPsc is resistant to proteases and accumulates in the lysosomes and eventually causes lysis of the cell.
		6. Cell lysis allows the PrPsc to spread to other neurons.
		7. The cell death results in the classic Swiss cheese appearance of the spongiform encephalitides.
			Spongiform encephalitis refers to sponge-like holes in the cerebral tissue.
			
			The pathology is remarkable for the degree of tissue destruction *without* inflammatory infiltration.
			
			PrPsc coalesces to form birefringent proteinaceous aggregates
			
			astrogliosis
			
			spinal fluid is normal
			
			Sterilization procedures which act to destroy nucleic acids are ineffective: ionizing radiation.
			
			To sterilize need to use method which destroys or denatures proteins.

Animal Prion diseases

Disease	Transmission	natural history	notes
Bovine Spongiform Encephalopathy (Mad Cow Disease)	Originally thought to come from cows ingesting cattle feed made from sheep brains affected with scrapie. Cattle feed then made with offals further spread the disease. **Special Bovine Offals** are cow parts which are unpalatable to humans and are made into animal feed. Includes: spleen, tonsils, thymus, brain and spinal cord .	occurs in cows causes the cows to become uncoordinated and lose their balance Most epidemics have been in Europe and England, almost 200,000 affected.	Oprah described BSE as a medical mystery spreading panic across the Atlantic Major concern is whether BSE can be transmitted to humans. Their has been an increase in CJD since the BSE epidemic in Britain but no hard evidence of bovine to human transmission has been acquired.
Scrapie		occurs in sheep and goats causes loss of coordination to the point which affected animals are unable to stand. also causes skin irritability such that the animals will scratch their hair off their body, this lead to the name scrapie.	recognized 250 years ago
Transmissible mink encephalopathy chronic wasting of mule deer and elk Feline Spongiform Encephalopathy			

Human Prion diseases

Disease	Transmission	natural history	notes
Cruetzfeldt-Jakob Disease (CJD)	Sporadic: idiopathic or iatrogenic Iatrogenic transmission has been documented with: • durae mater grafting • corneal transplant • growth hormone derived from human pituitaries • inadequite sterilization of surgical equipment Inherited: Autosomal dominant • 10% of all cases • Mutation in the gene coding for PrPc Occurs world wide Incidence is 1 in 1,000,000	Typically strikes people over the age of 45. Peak incidence is in the 50's Can have a prolonged latent phase lasting anywhere from 8 years to decades. The symptoms are a rapidly progressive dementia with an associated myoclonus (myoclonus is rare in Alzeimers). The timing from the onset of symptoms to death typically (90%) lasts less than one year. CT and MRI show cerebral atrophy late in the disease course but can be normal early in disease. EEG is helpful in diagnosis.	In Britain their has been a recent out break of atypical CJD characterized by younger patients: all under 42 years. There is a concern that these atypical cases could be secondary to the BSE epidemic of 1992. Incidence is increasing in the US but the significance of this is unclear. Possibly is an artifact of increased awareness among doctors of the disease leading to increased reporting.
Kuru	Cannibalism of affected human brains. Was more frequent among women who participated in the ritual of eating the brains of the deceased. Before prohibiting cannibalism the incidence was 1 in 100 members of the tribe.	Affected only the Fore tribe in Papua New Guinea Causes ataxia, dementia, intension tremors, myoclonic jerks, and choreoathetoid movements. Death within 1 year	
Gerstmann-Straussler-Scheinker disease	inherited condition	ataxia	
Fatal familial insomnia	inherited condition	dementia follows difficulty sleeping	

Prions

RNA Viruses

Section three: Fungi

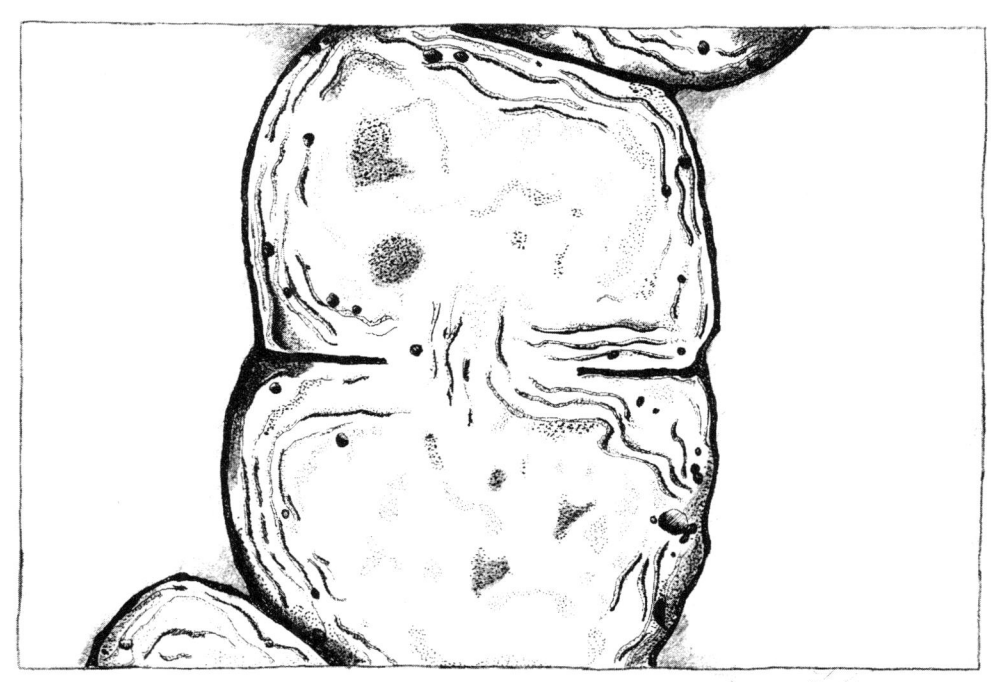

hyphae, Dennis Franklin

Fungi: Yeasts and Molds

Fungi, unlike bacteria, are **eukaryotic** organisms. Most are strict **aerobes**.

characteristic	bacteria	fungi
cell wall	peptidoglycan, glycerol and LPS	peptidomannan, glucan and chitin
plasma membrane	no cholesterol, except for *Mycoplasma*	contains a cholesterol called ergosterol which is not found in animal cells, targeted by some antifungal drugs
reproduction	fission, rare genetic recombination	asexual budding and sexual reproduction with meiosis
genome	single chromosome	multiple chromosomes, diploid during sexual reproduction
mobility	variable	immobile
organelles	none	mitochondria, ER, etc.

◆ Fungi can exist in two forms:
> **Yeasts** are single cells which reproduce by asexual budding.
> **Molds** are multinucleated and grow as long ribbon like strands known as **hyphae**.
>> **Hyphae** come in two varieties:
>>> **Septate** hyphae have walls dividing the hyphae into individual cells.
>>> **Nonseptate** hyphae are not divided, each hyphae is a large multinucleated cell.
> **Dimorphic** fungi grow as yeasts or molds depending on the environment. As a general rule they exist as **yeasts in the body** and **molds outside the body**.

◆ Fungi can reproduce sexually or asexually. Sexual reproduction occurs after mating and production of sexual spores. Asexual reproduction occurs by the formation of asexual spores known as **conidia**.

◆ Infections caused by fungi are often chronic and trigger **cell mediated immunity**, hence granuloma formation and delayed type hypersensitivity skin testing are common aspects of fungal infection.

◆ Treatment options for diseases caused by fungi are limited. Amphotericin B is most commonly used for the most serious infections, but also has the most serious side effects (hence the nickname: ampho-terrible). Other choices include: nystatin; the azoles (e.g. ketoconazole, fluconazole, etc.), flucytosine and griseofulvin.
> ⊃ Please see *The Pharmacology Companion*, pages 185 - 189 for more details.

◆ The fungi are grouped according to infection type: **systemic**, **opportunistic** and **cutaneous**.

systemic	opportunistic	cutaneous
All systemic fungi are dimorphic.	No opportunistic fungi is dimorphic.	All cutaneous fungi are dimorphic.
Coccidioides immitis	*Candida albicans* [yeast]	*Dermatophytes*
Blastomyces dermatitidis	*Cryptococcus neoformans* [yeast]	*Epidermophyton*
Histoplasma capsulatum	*Aspergillus fumigatus* [mold]	*Microsporum*
Paracoccidioides brasiliensis	*Zygomycetes* [mold]	*Trichophyton*
	Mucor	*Malassezia furfur*
	Rhizopus	*Cladosporium werneckii*
		Sporothrix schenckii

pathogen	lab findings	diseases	notes
Coccidioides immitis	**Dimorphic** fungus exists as a mold in soil and spherules in lungs. Should not be cultured because of the danger to lab workers. Diagnosis can be made by observation of spherules on KOH mount. Spherules are large (12-100 microns) with thick coats. Endospores can be seen inside. Skin test is positive 2-4 weeks after infection. Skin test is negative in persons with disseminated disease.	**Coccidioidomycosis** is endemic in the southwest U.S. and Latin America. **Transmitted** by inhaling arthrospores (syn. arthroconidia) contained in soil and dust. Arthrospores infect the lungs and form **spherules** which are structures containing hundreds of endospores. When spherules mature, they rupture and release endospores. At this point, if the body has developed T-cell immunity, activated macrophages destroy the endospores ending the infection. Most patients are asymptomatic. Inadequate cell-mediated immunity (i.e. AIDS, organ transplant, leukemia) allows **dissemination** throughout the body via the blood or direct extension within the lung. **Meningitis** is a serious complication. **Valley fever** is a hypersensitivity reaction which occurs in about 10% of infections. It is characterized by development of arthralgia, cough and fever. Erythema nodosum occurs in 50% of women. Lasts 4-6 weeks.	Treat with fluconazole. The spherule has antiphagocytic properties.
Blastomyces dermatitidis	**Dimorphic** fungus: **Mold** is found in the soil and grows in culture at 22-25°C. Mold has septate hyphae and round conidia. **Yeast** is found in the body and grows in culture at 37°C. The yeast is pear shaped, has a thick doubly refractive wall and a **broad based bud**.	**Blastomycosis** (syn. Gilchrist's disease) is endemic in southeastern U.S. including parts of the midwest. **Transmission** is by inhalation of conidia. Primary disease is **pulmonary** which can masquerade as TB, histoplasmosis or cancer. If the disease is unresolved, it can progress to **lobar pneumonia** and systemic dissemination via the blood. **Disseminated infection** causes ulcerated granulomas in the skin and bones. These will enlarge over years and can disfigure a large portion of the body.	Treat with itraconazole.

Systemic Mycoses

The Microbiology Companion, Topf and Faubel ©1997

167

Fungi

The Microbiology Companion, Topf and Faubel ©1997

Fungi

pathogen	lab findings	diseases	notes
Histoplasma capsulatum don't be fooled by the name, there isn't a capsule	**Dimorphic** fungus: **Mold** is found in soil and grows in culture at 22-25°C. It contains **septate hyphae**. Produces two types of sexual spores: **microconidia**: small and infectious **tuberculate macroconidia**: too large to infect humans. **Yeast** is found in the body and grows on Sabouraud's agar or BAP at 37°C. no capsule	**Histoplasmosis** is endemic in the central and eastern states and is prevalent in the Mississippi and Ohio River Valleys. Histoplasmosis shows many similarities to TB (the similarities will be underlined): **Transmission** is by inhalation of the microconidia. The body responds to infection by forming **caseous granulomas** in the lung which heal by calcification. The yeast is capable of multiplying inside macrophages and is spread throughout the body within these cells. **Diagnosis** of past infection can be done by a delayed hypersensitivity skin test. **TB symptoms** occur in a minority of cases when the initial lesion continues to grow and causes productive cough, fever and weight loss. **Asymptomatic infection** occurs in the vast majority of cases as demonstrated by the fact that 50% of the population in endemic areas has a positive skin test. **Disseminated infection** may occur in young, old and immunocompromised persons. The disease attacks the CNS, GI tract and adrenal glands but preferentially attacks the RES. **Reactivation** of latent infection may occur years after the primary infection when the individual becomes immunocompromised.	Treat moderate infection with itraconazole. Serious infections or infection in immunocompromised should be treated with amphotericin B. Treat with IV amphotericin B or ketoconazole. *Histoplasma* is found in bird droppings and **bat guano**. The pulmonary lesion which remains after healing appears as a discreet nodule on x-ray. The lesion is a **histoplasmoma** and can be mistaken as lung cancer. Diagnosis is by urinary ELISA which tests for *Histoplasma* antigen.
Paracoccidioides brasiliensis	**Dimorphic** fungus: **Mold** is found in the soil and appears filamentous. **Yeast** is found in the body and grows on BAP at 37°C. When reproducing the yeast has multiple buds. This is unique and useful for diagnosis.	**Paracoccidioidomycosis** is endemic in Central and South America. **Transmission** is by inhalation of conidia. Most infections are asymptomatic. **Disseminated infection** is initially characterized by ulcerative lesions in the oral and nasal mucosa. The organism infects local lymph nodes which ulcerate and drain through the skin. Other RES tissues can occasionally be infected.	The disease almost exclusively attacks males, possibly due to the protective effect of estrogen. Treat with itraconazole or ketoconazole.

Fungi

pathogen	lab findings	diseases	notes
Candida albicans	Only exists as a **yeast**. Reproduces by budding. In tissue, *Candida* may appear as budding yeasts or elongated growths which look like hyphae: **pseudohyphae** (true hyphae are only found in molds). Produce asexual **chlamydospores**.	**Candidiasis** may occur under the following circumstances: • **Lack of normal bacterial flora** usually occurs in patients on antibiotics or in newborns who have yet to acquire normal flora. • **Immune deficiency** such as AIDS, steroid therapy or leukemia • **Riboflavin deficiency** Diagnosis is by observing budding yeasts, pseudohyphae and chlamydospores from tissues or culture. Cultures can be deceiving because *Candida* is part of the normal flora. Candida infections may occur throughout the body: **Mouth** infection, known as:**thrush**, is characterized by white patches on the buccal membranes of the mouth. The majority of cases occur in infants and AIDS patients. It is common in the first 3 days of life because neonates can acquire the disease during passage through the birth canal. **Esophagitis** can occur with or without thrush. It causes substernal pain and painful swallowing, can cause perforation. **Skin** infection, **diaper rash** is a common example, occurs in warm and wet skin folds. Most common in infants, obese and diabetics. **Systemic candidiasis** is usually secondary to an implanted medical device, such as IV catheters and artificial valves. **Vaginitis** occurs most commonly in women who are diabetic, pregnant, on antibiotics or using oral contraceptives. Characterized by a cottage cheese-like vaginal discharge, inflammation around the genitalia, and intense itching. Laundering does not kill *Candida* so reinfection can occur from underwear.	Candida is part of the **normal flora** of the vagina, gut and oral cavity. Systemic infections should be treated with **amphotericin B**.

Opportunistic Mycoses

Fungi

pathogen	lab findings	diseases	notes
Cryptococcus neoformans	Only exists as yeast. **Capsule** is a polysaccharide containing mannose. 4 serotypes exist based on antigenic differences in the capsule. Urease + **India ink test** allows the organism to be visualized. The capsule excludes the ink allowing excellent contrast. Latex agglutination by anti-capsular antibodies aids diagnosis.	**Cryptococcosis** **Transmission** is by **inhalation** of conidia found in soil containing pigeon droppings. In immunocompetent people the yeast is phagocytized by PMNs. It is an opportunistic infection in patients with deficient cell mediated immunity (i.e. AIDS, steroid therapy). Infection is characterized by a **lack of tissue response**. A biopsy of the infected tissue will show yeast cells in the tissue with no inflammatory cells attacking them. **Meningitis** has a slow onset of non-specific symptoms: headache, irritability, dizziness, behavior changes, fever and/or seizures. Untreated the disease is 100% fatal but most cases can be controlled with antifungals. Relapses are common and permanent damage occurs in half of all cases.	This opportunistic infection often kills persons with AIDS. Can be isolated from 15% of people with AIDS. Think: puts people with AIDS in the CRYPT. Treat with amphotericin B and flucytosine.

Fungi

pathogen	lab findings	diseases	notes
Aspergillus fumigatus	Exists only as a **mold**. Tissue biopsy will show branching, **septate** hyphae. Produce asexual conidia. Thermophilic: grows at 45°C. Very fast growing. The molds are common in the environment and are found worldwide. The organism is so common that it is difficult to distinguish actual infection from laboratory contamination.	**Lung infections** occur by inhalation of airborne conidia or colonization of the lung. **Allergic bronchopulmonary aspergillosis** is characterized by eosinophilia, antibodies against aspergillus (IgG and IgE) and transient exudates. The production of antibodies results from chronic colonization and continual shedding of antigens. The resulting allergies can cause asthma and complicate other chronic lung diseases. **Farmer's lung** is a type of allergic bronchopulmonary disease which occurs after inhaling a large number of *Aspergillus* conidia found in hay silos. **Fungus balls** occur when the organism grows in cavities left by pulmonary disease (e.g. tuberculosis). Proliferation of the organism results in a fungus ball, which is visible on an x-ray. The growth may invade surrounding tissues causing hemoptysis and dissemination. **Disseminated infection** can result in the formation of **granulomas** in many tissues throughout the body.	Treat with amphotericin B for disseminated infection. Fungus balls must be surgically removed.
Zygomycetes *Mucor* *Rhizopus*	Saprophytic **molds** found in soil. No yeast form. Tissue biopsy will show **nonseptate**, broad hyphae.	**Zygomycosis** (syn. mucormycosis or phycomycosis) is caused by members of the class Zygomycetes such as *Mucor* or *Rhizopus*. These organisms have a propensity to proliferate in the walls of blood vessels. Immunocompromised persons are at risk for infection as are those with **diabetic ketoacidosis** (a complication of insulin dependent diabetes). **Pulmonary infection** may cause pneumonia which is clinically similar to other fungal pneumonias. **Rhinocerebral disease** occurs after inhalation of the organism which invades the nasal mucosa and proliferates in the blood vessels. The organism continues to invade and extends upward through all tissues until eventually reaching the brain. After the initial presenting symptom of headache, the disease can be fatal in two weeks.	The disease is treated by surgical removal of necrotic and infected tissue and administration of amphotericin B.

The Microbiology Companion, Topf and Faubel ©1997

Fungi

Cutaneous Mycoses

pathogen	lab findings	diseases
Dermatophytes are a classification of fungi which cause skin infections in humans, included are the following genera: *Epidermophyton* *Microsporum* *Trichophyton*	Produce asexual conidia Grow on Sabouraud's media.	The organisms require keratin for growth, therefore infections only occur in superficial, keratinized structures: skin, nails and hair. Subcutaneous infection does not occur. Infection begins in the skin and spreads in a radial fashion, producing a ring of inflammation. So although these infections have nothing to do with worms, they are called **ringworm** because of their appearance on the skin. The infections are named according to their location on the body and may be caused by any of the three genera of Dermatophytes. These infections include: *Tinea pedis* (athletes foot) *Tinea cruris* (jock itch) *Tinea corporis* (ringworm of the body) *Tinea capitis* (ringworm of the scalp) *Tinea unguium* (nail infection) *Tinea barbae* (hair infection) **Treatment** is with topical agents such salicylic acid, topical antifungal agents such as miconazole, and/or oral griseofulvin. **Griseofulvin** is incorporated into keratinized tissue and imparts resistance to fungal infection. **Dermatophytid** (usually abbreviated "id") **reactions** may occur in chronically infected persons. They are characterized by the appearance of vesicles and occur due to the development of hypersensitivity and subsequent allergic response to the fungus.

pathogen	lab findings	diseases	notes
Malassezia furfur	Dimorphic fungus. In culture, grows primarily as a yeast.	**Tinea versicolor** is characterized by <u>light</u> <u>lesions</u> which are hypopigmented areas which may appear yellow or pink. **Diagnosis** is by observation of budding yeast cells and hyphae together in KOH prepared skin scrapings.	Recurrences are frequent, even with treatment.
Cladosporium werneckii	Dimorphic fungus. Brown/black pigmented organisms found in soil. septate hyphae	**Tinea nigra** is characterized by <u>dark</u> <u>lesions</u> (due to color of the organism) on the skin, usually the hands and feet. The disease is most common in the tropics, but may be found in the U.S. Transmission occurs during skin injury. **Diagnosis** is by microscopic appearance of KOH prepared skin scrapings Infection occurs in southern states.	
Sporothrix schenckii	Dimorphic fungus. **Molds** are saprophytes found in soil and on plant surfaces. Characterized by the appearance of thin, septate hyphae. **Yeast** are found in tissue and grow in culture at 37°C. ubiquitous Produce asexual conidia.	**Sporotrichosis** is a <u>subcutaneous</u> infection. **Transmission** is commonly caused by being pricked by a **thorn** contaminated with *S. schenckii* spores. Gardeners and farmers are at risk for this infection. The organism multiplies under the skin and forms a pustule which eventually ulcerates. Sites along lymphatic drainage may also become infected. Without treatment, the lesions may become chronic. **Diagnosis** is by noting cigar shaped yeasts in a tissue biopsy and by identification of the organism in culture media. Rarely, the organism may spread systemically and deposit in the CNS, bones and/or lungs.	Treatment of cutaneous infection is with <u>oral</u> **potassium iodide**.

Section four: Parasites

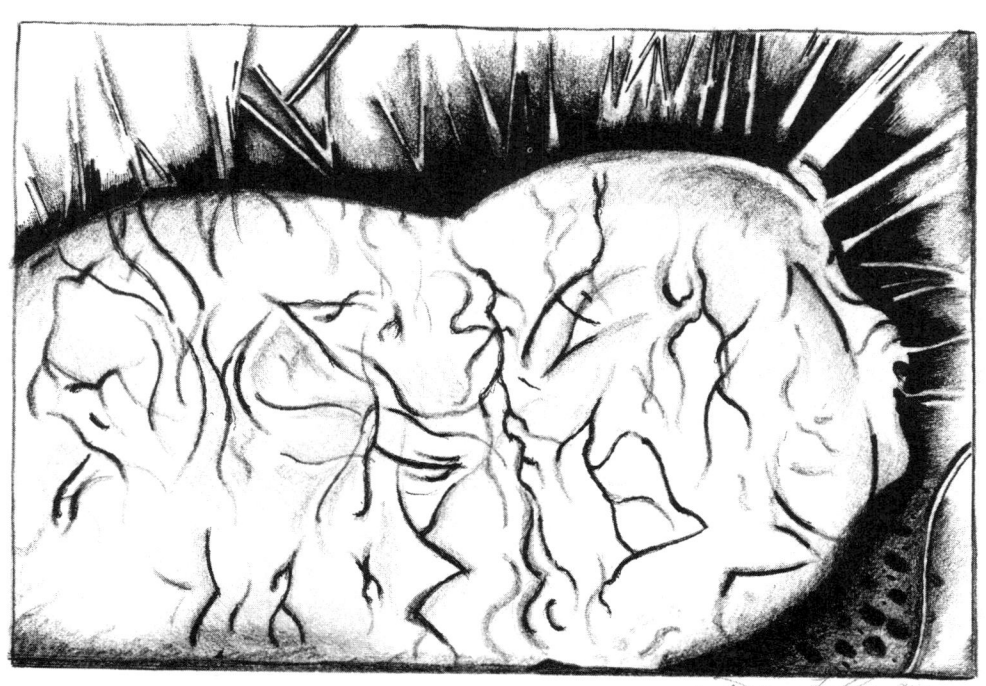

Proteus mirabalis, Dennis Franklin

Parasites, protozoa   The Microbiology Companion, Topf and Faubel ©1997   **178**

Parasites

Parasites are divided into two different classes: protozoa (unicellular) and helminths (multicellular worms). There are a wide variety of survival strategies employed by these tiny free loaders as they move between their multiple hosts.

Protozoa are not animals, plants or bacteria. They are eukaryotic organisms which comprise their own kingdom. Protozoa are unicellular parasites with a membrane around their nucleus and cytoplasm. The protozoa are able to exist in two different forms: trophozoites and cysts. **Trophozoites** are the motile form and **cysts** are the durable form. Cysts can survive outside the host and are able to withstand heat, dehydration and many chemicals (chlorination).

Three major classes of protozoa cause disease in humans:

Ameba	**Flagellates**	**Sporozoa**
Entamoeba histolytica	Intestinal flagellates	*Cryptosporidium*
	Giardia lamblia	*Plasmodium*
	Trichomonas vaginalis	*falciparum*
	Hemoflagellates	*vivax*
	Leishmania	*ovale*
	donovani	*malariae*
	braziliensis	*Toxoplasma gondii*
	mexicana	
	tropica	
	Trypanosoma	
	cruzi	
	brucei	
	gambiense	
	rhodesiense	

Pneumocystis carinii is considered a protozoa for medical purposes even though genetic tests have determined it to be a fungus.

Ameba are a subclassification of protozoa which are able to send out pseudopodia from the cell to assist in movement and engulfing food. They are unable to reproduce sexually.

pathogen	appearance	diseases and life cycle	important notes
Entamoeba histolytica	The protozoa exists in two forms: **Trophozoite** has a single nucleus and no flagella. It moves by extending pseudopods and filling them with cytoplasm. Trophozoites can be identified by the presence of RBCs inside them because *Entamoeba histolytica* is the only protozoa which dines on them. Reproduce by fission. The trophozoite can not survive outside the digestive tract. **Cysts** are formed when the trophozoite is pulled off the intestinal wall by fecal matter moving down the digestive tract. To become a cyst, the trophozoite excretes unessential organelles and shrinks while dividing its nuclei twice to produce <u>four nuclei</u>. The ribosomes aggregate into dark bars known as **chromatoidal bars**, which are normally found in pre-cysts and occasionally seen in the mature cysts.	**Amebiasis** is colonization of the human intestine with or without disease. In asymptomatic cases, the ameba feeds on intestinal flora. This occurs in 90% of amebiasis. **Amebic dysentery** is amebiasis with disease. This occurs when the ameba invade the intestinal walls to **feed on RBCs**. The intestinal wall is eroded by the amebas. Rarely the amebas will eat their way into the portal circulation. When this occurs, they are able to travel into the liver and form abscesses. **Transmission** occurs by the **fecal-oral** route and is associated with poor sanitation. <u>The disease is spread only by the cyst form</u> which is excreted by asymptomatic carriers. Symptomatic individuals do not spread the disease because they shed trophozoites which are not infectious (too fragile to exist outside the body).	Treat amebic dysentery with metronidazole. Common among homosexual men due to oral-anal contact.

Flagellates are a group of protozoa which are divided into two groups: intestinal and hemoflagellates. The intestinal flagellates are non-invasive and cause relatively benign intestinal and genital tract infections; they do not have intermediate hosts (insect vectors). The hemoflagellates cause invasive infections in the blood which can be very severe and are transmitted by insects (arthropods).

Intestinal Flagellates

◆ non-invasive, infect the lumen of the GI and GU tract, no intermediate host

pathogen	appearance	diseases and life cycle	important notes
Giardia lamblia	This flagellate protozoa exists in two forms: **Trophozoites** are bilaterally symmetric (the left and right sides are mirror images) and shaped like a pair. They have two nuclei, four pairs of flagella and a sucker which allows attachment to the intestinal wall. **Cysts** are oval shaped with a thick protective wall and four nuclei. One cyst is able to produce two trophozoites. Cysts can survive chlorine treatment so drinking water should be treated with boiling, iodine or filtering.	**Giardiasis** is an intestinal infection characterized by smelly, non-bloody diarrhea which persists for weeks. Other symptoms include abdominal cramps, nausea and flatulence. **Transmission** occurs by the ingestion of cysts. Cysts are found in fecally contaminated water. After ingestion, the cysts undergo excystation in the duodenum and produce two trophozoites. Trophozoites attach to the duodenal wall with the sucker and cause inflammation and rarely malabsorption of fat and protein. Infection by *Giardia* is the most frequent cause of water born diarrhea in the U.S. and the **most common intestinal parasite.**	Treat with metronidazole. 5% of the population are asymptomatic carriers excreting cysts in their stool. IgA deficiency increases chance of symptomatic disease. The disease is common among **hikers** who may drink untreated water. The infection is also common in **homosexual males** due to anal-oral contact.

pathogen	appearance	diseases and life cycle	important notes
Trichomonas vaginalis (this is not the same as *Chlamydia trachoma*)	This flagellate protozoa **exists only as a trophozoite**, there is no cyst form. The trophozoite is pear-shaped with a central nucleus and **4 external flagella.**	**Trichomoniasis** is a common **sexually transmitted disease** (STD). Symptoms in women include burning, itching and a copious, foul smelling vaginal discharge which may appear yellow, green and/or bubbly. Loss of the normal acidity of the vagina predisposes to infection. Men are usually asymptomatic, but some have urethritis. Infection usually occurs in the prostate. **Diagnosis** is by observation of a <u>wet mount</u> of vaginal or prostatic secretions which will show trophozoites moving on the slide.	Treat both partners with metronidazole. Up to 25% of sexually active women in the U.S. will be infected with *Trichomonas vaginalis* at some point in their lives.

Hemoflagellates

◆ arthropod vector, intermediate host, invasive infection

pathogen	transmission	diseases
Leishmania donovani	**Sandfly** is the vector and humans and animals are reservoirs. Exists as two forms: **Promastigotes** are the flagellated form found in the gut and pharynx of the sandfly. **Amastigotes** are the nonflagellated form found <u>intracellularly within macrophages</u> of the reticuloendothelial system in a human or animal host. **Life cycle**: Promastigotes are injected into an animal or human host when a female sandfly sucks out blood. After engulfment by macrophages, the promastigotes revert to the amastigote form. The amastigotes multiply within macrophages which eventually die and release more amastigotes which infect more macrophages. Another sandfly comes along and sucks blood containing amastigotes within the macrophages. The amastigotes become promastigotes in the gut and then migrate and multiply within the pharynx from whence they can be injected into another unsuspecting host.	**Kala-azar** (**visceral leishmaniasis**) attacks the reticuloendothelial system (fixed phagocytic cells) of the skin, spleen, liver and bone. The abdomen may swell due to the enlargement of the spleen and liver secondary to the proliferation of organisms inside these organs. **Intracellular growth** within macrophages occurs when the organism is in the amastigote (nonflagellated) form. **Diagnosis** is by demonstration of parasites in a spleen or liver biopsy. It was feared that military personnel involved in Desert Storm might contract this infection and bring it to the U.S.
Leishmania braziliensis mexicana tropica	**Sandfly** is the vector and humans and animals are reservoirs.	**Cutaneous leishmaniasis** occurs when the organism only invades the reticuloendothelial system (fixed phagocytic cells) of the **skin**. A <u>granulomatous</u> lesion forms at the site of the insect bite secondary to induction of cell mediated immunity. The infection may disseminate and cause widespread skin and organ lesions in persons with depressed cellular immunity.

Notes

Parasites

Protozoa, Hemoflagellates, *Trypanosoma*   The Microbiology Companion, Topf and Faubel ©1997   **184**

pathogen	transmission	diseases
Trypanosoma cruzi	**Reduviid bug** is the vector and humans and animals are intermediate hosts. Trypanosomes grow in the bug gut and are passed in its feces. The reduviid bug typically defecates while biting, so **scratching** by the host improves transmission of the organism.	**Chagas' disease** (American trypanosomiasis) occurs in Central and South America where it is the leading cause of heart disease. **Transmission**: the reduviid bug hides in thatch walls of houses and comes out at night to feed on people while they are sleeping. Its favorite spot to bite seems to be around the eyes and mouth which has earned it the title of "kissing bug." Organisms multiply at the site of the bite and a chancre (chagoma) forms. Dissemination occurs through the blood and the organism deposits throughout the body, particularly the **brain**, glial cells and **heart**, **skeletal** and **smooth muscle**. **Diagnosis** is by examination of a peripheral blood smear for the presence of trypomastigotes. Most infections are asymptomatic. **Acute infection** usually occurs in children and is characterized by fever, lymphadenopathy, and hepatosplenomegaly. **Chronic disease** may occur in individuals with a symptomatic or asymptomatic primary infection. Infection may cause a continuous inflammatory response leading to tissue destruction. Damage to the **heart** is the most serious complication and may present as arrhythmia or congestive heart failure. Neurons of the GI tract may be affected and cause loss of peristalsis.

pathogen	characteristics	diseases
Trypanosoma brucei:; has three subspecies: *brucei* *gambiense* *rhodesiense*	*Trypanosoma gambiense* causes West African sleeping sickness and *Trypanosoma rhodesiense* causes East African sleeping sickness. Although the diseases are distinctly different, *T. gambiense* and *T. rhodesiense* are morphologically and antigenically indistinguishable. *Trypanosoma brucei* is not known to cause any human infection, but it is also morphologically and antigenically indistinct from *T. gambiense* and *T. rhodesiense*. The three Trypanosomes are thus considered to be variations of the single species *Trypanosoma brucei*. The three subspecies are distinguished by biologic characteristics and are known as *Trypanosoma brucei gambiense*, *Trypanosoma brucei rhodesiense* and *Trypanosoma brucei brucei*.	**Sleeping sickness** (African trypanosomiasis) **Transmission** occurs during the bite of an infected **tsetse fly** when trypomastigotes are injected into the skin. A skin ulcer (chancre) forms at the site of the bite. Trypomastigotes spread through the blood and lymph nodes to eventually reach the brain. Initial symptoms of fever, rash, and lymphadenopathy occur 2-3 weeks after being bit. Trypanosomes exhibit **antigenic variation** of a surface glycoprotein, known as variable surface glycoprotein (VSG), which corresponds with the characteristic **return of fever and other symptoms every two weeks.** The organism causes meningoencephalitis which progresses from headache to muscle tremors, slurred speech, difficulty walking, decreased attention, and ends with seizures, somnolence and coma. Untreated disease is usually fatal. **Diagnosis** is by microscopic examination of thick and thin blood smears for the presence of trypomastigotes. CSF may also be examined. Serologic tests for IgM antibodies are also useful in diagnosis.
Trypanosoma brucei gambiense	**Tsetse fly** is the vector and **humans** are the major reservoir. *T. gambiense* is spread by a different species of tsetse fly than *T. rhodesiense*.	**West African Sleeping sickness** (African trypanosomiasis) is caused by tsetse flies indigenous to West and Central Africa, particularly near water sources. The disease runs a low grade chronic course which can run for several years before clinical evidence of CNS involvement.
Trypanosoma brucei rhodesiense	**Tsetse fly** is the vector and animals, particularly **antelopes**, are the major reservoir. *T. rhodesiense* is spread by a different species of tsetse fly than *T. gambiense*.	**East African Sleeping sickness** (African trypanosomiasis) is caused by tsetse flies indigenous to East and Central Africa, particularly arid regions. The disease is **more severe** than the West African variety. It progresses rapidly and is often fatal before the appearance of CNS symptoms. Infection of the **heart and CNS** occur within 3-6 weeks. Death occurs within 6-9 months from heart failure or coma.

Sporozoans

Sporozoa are the only medically important protozoa to undergo sexual reproduction. They also maintain the ability for asexual reproduction. Asexual reproduction does not occur by fission (two equal daughter cells from a single mother cell). The sporozoa reproduce through a process known as schizogony in which the nucleus undergoes multiple divisions and forms a multinucleated *schizont*; then cytoplasm surrounds each daughter nucleus and the mother cell lysis releasing multiple new *merozoites*. During sexual reproduction, called sporogony, some merozoites differentiate into male and female *gametocytes*, which join to form a zygote. Zygotes matures into oocysts which hatch multiple asexual sporozoites that mature into the adult asexual trophozoite.

Sporozoa

Cryptosporidium

Plasmodium

 falciparum

 vivax

 ovale

 malariae

Toxoplasma gondii

pathogen	appearance	diseases and life cycle	important notes
Cryptosporidium	**Oocysts** are the infectious form. **Trophozoites** are formed in the intestines of the host.	**Cryptosporidiosis** is an intestinal infection characterized by watery, non-bloody diarrhea which occurs particularly in immunocompromised persons. **Transmission** is by the **fecal-oral** route. The trophozoite form of the parasite attaches to the gut wall of the ileum but does not invade. **Chronic diarrhea** occurs in persons with AIDS. **Self-limited diarrhea** occurs in the immunocompetent.	Treatment is supportive, with replacement of fluids and nutrients. Common infection in persons with AIDS.

Plasmodium causes malaria

◆ Sexual reproduction (sporogony) occurs in the gut of mosquitoes.

◆ Asexual reproduction (shizogony) occurs in vertebrate hosts.

◆ Symptoms, such as fever, relapse every 2 or 3 days.

◆ *P. vivax* and *P. ovale* can induce a latent infection of the liver which can last years before a relapse.

pathogen	disease
Plasmodium *falciparum* *vivax* *ovale* *malariae*	**Malaria** is the **most common lethal disease** in the world, killing over one million people a year. **Transmission** occurs after the bite of an infected female *Anopheles* mosquito, which is both the vector and definitive host for *Plasmodium*. The disease is characterized by sudden onset of fever, chills and sweating about two weeks after the mosquito bite. Headache, myalgia, arthritis, nausea, vomiting, abdominal pain, splenomegaly, hepatomegaly and/or anemia may also occur. A recurrence of symptoms occurs every 48 to 72 hours (depending on the species, see below) when merozoites are released into the circulation and infect more red blood cells. **Diagnosis** is by microscopic examination of blood with thick and thin smears stained with Giemsa or Wright stain. **Thick smears** are necessary to determine the presence of parasites (the chances of actually seeing parasites are higher when the RBCs are concentrated as in a thick stain). **Thin smears** are used to identify the particular type of *Plasmodium* infection.

Plasmodium	characteristics	type of malaria
falciparum	Causes the most severe form of malaria infection. Infects the most RBCs. Infects RBCs of all ages; very severe infections can occur.	**Malignant tertian malaria**: fever recurs every 48 hours **Blackwater fever**: infection with *P. falciparum* causes the destruction of a large number of RBCs which can occlude capillaries and cause tissue destruction. When kidney tissue is affected, dark urine ("black water") is produced as a result of blood in the urine (hematuria). The disease is <u>potentially life-threatening</u> due to extensive kidney and liver damage.
vivax	Can produce a latent form of infection in the liver by the production of **hypnozoites**. Infects only reticulocytes (immature RBCs) so only a 1-2% of RBCs are susceptible.	**Benign tertian malaria:** fever recurs every 48 hours. Self-limited, low mortality.
ovale	Can produce a latent form of infection in the liver by the production of **hypnozoites**. Infects only reticulocytes (immature RBCs) so only a 1-2% of RBC's are susceptible.	**Benign tertian malaria:** fever recurs every 48 hours. Self-limited, low mortality.
malariae	Infects only old RBCs so only a 1-2% of RBC's are susceptible.	**Quartan malaria:** fever recurs every 72 hours. Self-limited, low mortality.

A common question about malaria refers to its association with sickle cell trait. Areas in which malaria is endemic have higher rates of sickle cell anemia. This is because the sickle cell trait (Hgb AS, not the SS form which is the very serious disease sickle cell anemia) is protective. Many explanations have been given for this; check off the one given in your class and ignore the others:

① Since the life span of RBCs with the sickle trait is lower the parasite is unable to complete its life cycle before the cell is cleared by the spleen.

② The increased turnover of RBCs decreases the available folate (a metabolite necessary for DNA and RNA synthesis) preventing the parasite from reproducing efficiently.

③ The sickle hemoglobin causes direct damage to the parasite.

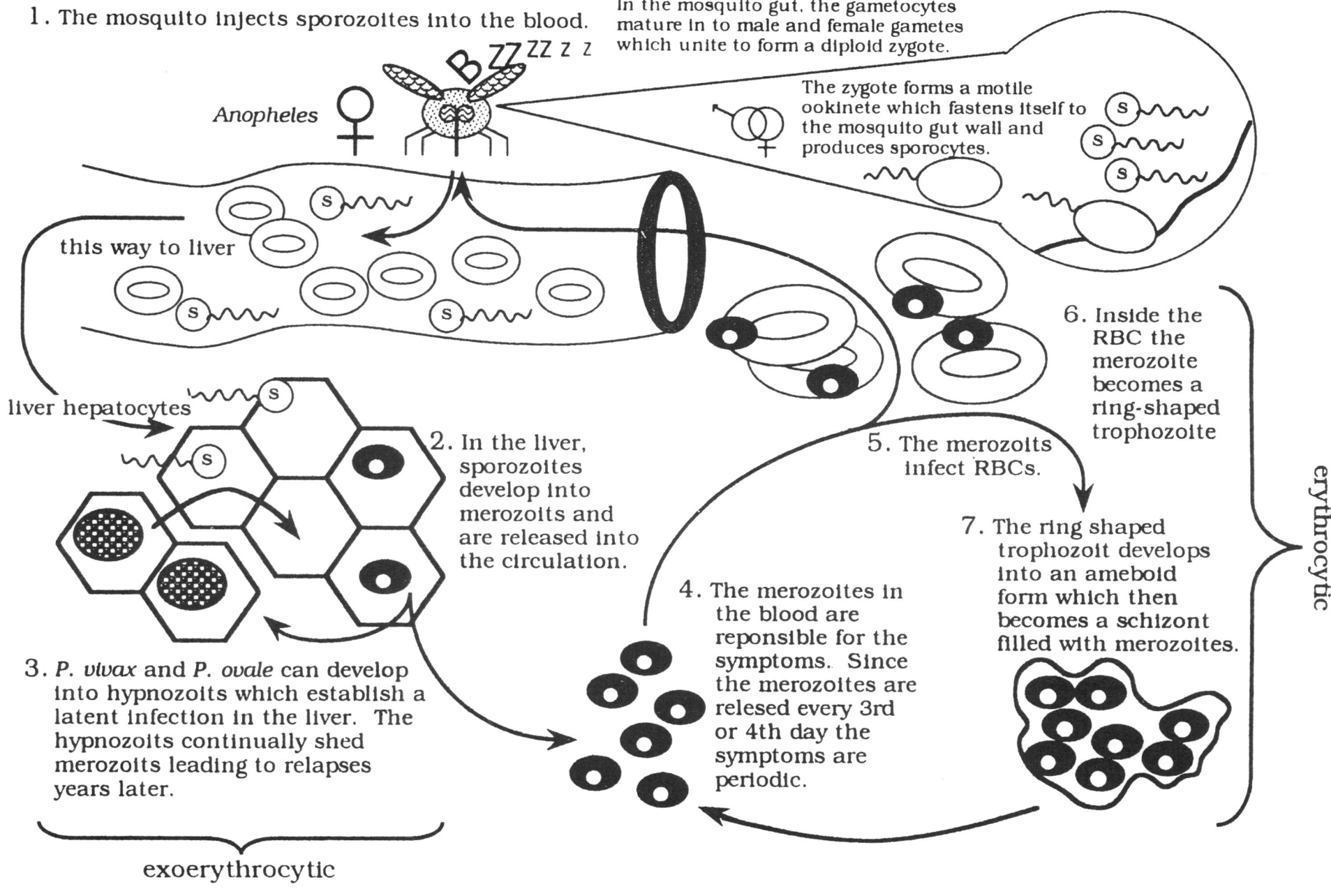

1. The mosquito injects sporozoites into the blood.

In the mosquito gut, the gametocytes mature in to male and female gametes which unite to form a diploid zygote.

The zygote forms a motile ookinete which fastens itself to the mosquito gut wall and produces sporocytes.

Anopheles ♀

this way to liver

liver hepatocytes

2. In the liver, sporozoites develop into merozoits and are released into the circulation.

3. *P. vivax* and *P. ovale* can develop into hypnozoits which establish a latent infection in the liver. The hypnozoits continually shed merozoits leading to relapses years later.

exoerythrocytic

4. The merozoites in the blood are reponsible for the symptoms. Since the merozoites are relesed every 3rd or 4th day the symptoms are periodic.

5. The merozoits infect RBCs.

6. Inside the RBC the merozoite becomes a ring-shaped trophozoite

7. The ring shaped trophozoit develops into an ameboid form which then becomes a schizont filled with merozoites.

erythrocytic

Protozoa, Sporozoa, *Plasmodium*

Parasites

Protozoa, Sporozoa, *Toxoplasma* — The Microbiology Companion, Topf and Faubel ©1997 — **190**

pathogen name	disease	notes
Toxoplasma gondii	**Toxoplasmosis** is a common parasitic infection in the U.S. **Transmission** occurs through a variety of mechanisms: • **Ingestion** of cysts in raw or undercooked meat, particularly pork and mutton. • **Contact** with oocysts in cat feces. Persons changing kitty litter and children, who may play in areas contaminated with cat feces (sandboxes), are at risk of infection. • **Transplacental** to fetus during primary infection of the mother when circulating organisms cross the placenta. **Diagnosis** is by serologic detection of IgM antibodies or a four-fold increase of IgG. The type of infection produced depends on the type of host: **Immunocompetent adults** may present with symptoms resembling **mononucleosis**: lymphadenopathy, fever, and sore throat. **Immunocompromised persons** may have a severe, life threatening disease caused by **reactivation** of a primary infection. Reactivation usually occurs in the brain. **In utero infection** commonly causes abortion and still birth. Infants who survive may have microcephaly, hydrocephaly, and other CNS effects which result in mental retardation. Infants may be asymptomatic at birth only to develop symptoms of infection months or years later. A common delayed manifestation of congenital infection is **chorioretinitis** which causes vascular changes in the retina and may lead to blindness.	About half of the adult population in the U.S. have been infected. Toxoplasma is the most common CNS infection found in persons with AIDS. Treat with sulfonamide and pyrimethamine. **Cell mediated immunity** plays the principle role in limiting infection. The entire life cycle (sexual and asexual) occurs in the GI tract of **cats** which are the definitive host.

pathogen	diseases	important notes
Pneumocystis carinii The classification of *P. carinii* is unclear. Earlier it was thought that *Pneumocystis carinii* was a Trypanosome (a flagellate protozoa). Recent genetic analysis indicates that *Pneumocystis carinii* is actually a **fungus**. Paradoxically, antifungal medication is ineffective against the organism. Therefore **it is medically still considered a protozoa** and will be discussed here.	**Pneumocystis carinii pneumonia** (PCP) in immunocompromised persons. **Transmission** occurs by the respiratory route. **Symptoms** include sudden onset of fever, non-productive cough, shortness of breath, and rapid breathing. **Pathogenesis**: the organisms attaches to and destroys type I alveolar cells (the thin cells of the alveoli, the type II cells are the ones which produce surfactant); fluid leaks into the alveolar air spaces and causes hypoxemia. **Diagnosis** is by identifying the organism in sputum or lung tissue with special stains (Giemsa stain or Gram-Weigert stains). The increase in PCP has paralleled the incidence of AIDS and the use of immunosuppressive agents. PCP is the most common opportunistic infection found in AIDS patients in North America and Europe. PCP is uncommon in Africa where TB is the major opportunistic infection in AIDS patients. In the U.S. PCP is the AIDS defining illness in over half of all AIDS patients. in the U.S. up to 80% of AIDS patients will acquire PCP at some point. **Asymptomatic infection** is common. The organism is ubiquitous in the environment and distribution is worldwide. Studies in the U.S. have shown that circulating antibodies to *P. carinii* develop in most children by age 2 or 3.	**Treatment** of PCP is with IV trimethoprim and sulfamethoxazole. Prophylaxis is recommended for HIV+ persons with a CD+4 count of less than 200 or with a previous history of PCP. Agents used for prophylaxis include: 1. oral trimethoprim-sulfamethoxazole 2. oral dapsone 3. aerosolized pentamidine

Parasites

The Microbiology Companion, Topf and Faubel ©1997

Helminths (worms) are multicellular parasites which lack both a vascular and a respiratory system. Helminths are divided into two classes: Platyhelminths (flatworms) and Nematodes (roundworms). The Platyhelminthes (flatworms) include the Cestodes (tapeworms) and Trematodes (flukes). The Nematodes (roundworms) include intestinal and tissue types.

Platyhelminthes (flatworms)

Cestodes (tapeworms)

Taenia saginata

Taenia solium

Diphyllobothrium latum

Echinococcus granulosus

Trematodes (flukes)

Schistosoma

   mansoni

   japonicum

   hematobium

Clonorchis sinensis

Paragonimus westermani

Nematodes (roundworms)

Intestinal

Enterobius (pinworm)

Trichuris (whipworm)

Ascaris

Ancylostoma (hookworm)

Necator (hookworm)

Strongyloides

Tissue

Trichinella (trichinosis)

Ancylostoma larvae

Toxocara larvae

Wuchereria

Dracunculus (guinea worm)

Onchocerca (river blindness)

Brugia malayi

Loa loa

Helminth, Platyhelminth, Cestodes (tapeworms)   The Microbiology Companion, Topf and Faubel ©1997   **194**

Cestodes also known as **tapeworms** are a type of platyhelminth (flatworm).

Platyhelminthes (flatworms)

Cestodes (tapeworms)
Taenia saginata
Taenia solium
Diphyllobothrium latum
Echinococcus granulosus

Trematodes (flukes)
Schistosoma
   mansoni
   japonicum
   hematobium
Clonorchis sinensis
Paragonimus westermani

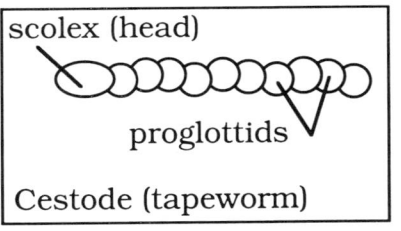

scolex (head)

proglottids

Cestode (tapeworm)

Cestodes consist of a scolex (head) and proglottids (segments). The scolex (head) contains some mechanism for attaching to the intestinal wall of the host: suckers, hooks, or sucking grooves. The most distal (at the tail end) proglottids are gravid: they produce eggs. The egg containing distal proglottids detach and are passed in the feces.

The eggs from contaminated feces are ingested by an intermediate host, such as a cow (*Taenia solium*) a pig (*Taenia saginata*) a fish (*Diphyllobothrium latum*) or a dog (*Echinococcus granulosus*). In the intermediate host, the eggs mature into larvae which deposit in muscle and/or other tissues of the animal. The definitive host, such as humans (*Taenia saginata*, *Taenia solium*, *Diphyllobothrium latum*) or sheep (*Echinococcus granulosus*), eat the infected tissue. In the definitive host, larvae attach to the intestine and grow into a mature adults which produce eggs.

Life cycle summary:

◆ The adult cestode (tapeworm) lives in the intestines of the host where eggs are laid.

◆ Maturation from eggs to larvae occurs in the intermediate host.

◆ Maturation from larvae to adult occurs in the definitive host.

Summary of cestodes (tapeworms)

Cestode (tapeworm)	definitive host	intermediate host(s)	disease
Taenia saginata	humans	cattle	usually asymptomatic
Taenia solium	humans	pigs	asymptomatic: from ingestion of larvae cysticercosis: from ingestion of eggs
Diphyllobothrium latum	humans, foxes, bears, cats (any mammal which eats raw fish)	copepod (a crustacean) (1st intermediate host) fish (2nd intermediate host)	usually asymptomatic deficiency of vitamin B_{12} megaloblastic anemia
Echinococcus granulosus	dogs	sheep	hydatid cyst disease

Treatment options for cestodes include: praziquantel, albendazole and niclosamide.

⊃ For a review of treatments, see *The Pharmacology Companion* pages 205 - 208.

Parasites

Helminth, Platyhelminth, Cestodes (tapeworms)

The Microbiology Companion, Topf and Faubel ©1997

pathogen	appearance	diseases and life cycle	important notes
Taenia saginata BEEF TAPEWORM	**scolex:** 4 suckers, no hooks **gravid proglottids**: contain 15-25 uterine branches (this feature is used to distinguish this tapeworm from others) **eggs**: **round**, morphologically identical to *T. saginata* and *Echinococcus*	**Life cycle** • **Eggs** are passed in feces and then ingested by *cattle* (the intermediate host). • In cattle, the eggs hatch in the intestine and the **embryos** travel by blood and lymphatics to eventually deposit in <u>striated muscle</u>. • In the muscle, the embryos develop into cysticerci, the larval form. A **cysticercus** is a fluid filled sac which contains an invaginated scolex. • Transmission to *humans* occurs by the ingestion of raw or undercooked <u>beef</u> containing larvae (cysticerci). The larvae attach to the intestine and grow into the **adult** form. • **Adults** may live in the intestines for as long as **25 years** and grow up to **12 meters**. Infection is usually asymptomatic. Some persons may have diarrhea, weight loss and/or nausea. <u>Does not cause cysticercosis.</u>	**Humans** are the only definitive hosts. **Cattle** are the intermediate hosts.

pathogen	appearance	diseases and life cycle	important notes
Taenia solium PORK TAPEWORM	**scolex**: 4 suckers and a ring of hooks **gravid proglottids**: contain 5-10 uterine branches **eggs**: **round**, morphologically identical to *T. saginata* and *Echinococcus*	The **life cycle** when larvae are ingested is the same as the life cycle for *T. saginata* (see above) except pigs are the intermediate host. The type of disease caused by infection is dependent on the form of the organism ingested: larvae or eggs. • Ingestion of **larvae** in undercooked **pork**. Infection is usually asymptomatic, but may cause diarrhea and/or weight loss. **Diagnosis** relies on finding eggs and proglottids in stool. • Ingestion of **eggs** in contaminated water or food causes **cysticercosis**. After the eggs hatch in the small intestine, the organism spreads throughout the body, deposits in various tissues and forms cysticerci (larval form). Symptoms usually appear years after the initial infection and when the organisms die and induce inflammatory reactions. The lesion calcifies and can be seen on x-ray. Preferential sites of infection are the **brain** and **eyes**.	**Humans** are the only definitive hosts. **Pigs** are the intermediate hosts.

Parasites

Helminth, Platyhelminth, Cestodes (tapeworms) The Microbiology Companion, Topf and Faubel ©1997 **198**

pathogen	appearance	life cycle	diseases
Diphyllobothrium latum FISH TAPEWORM	**scolex**: 2 sucking grooves, no hooks **gravid proglottids**: uterus forms a rosette **eggs**: **oval** with an operculum (lid-like opening) **Diagnosis** relies on finding eggs and proglottids in stool.	• **Adults** live in the intestine of a *mammalian host* and produce **eggs**. • Eggs develop into free-swimming **coracidia**, which are consumed by *copepods* (a type of crustacean). • In the copepod, the coracidium becomes a **procercoid**. After the copepod is ingested by a *fish* the procercoid becomes a **plerocercoid** and imbeds itself in the muscle of the fish. • The cycle is completed when a *mammalian host* ingests raw fish containing plerocercoids. **Humans** and other raw fish eating mammals (foxes, bears, cats) are the definitive hosts. The **copepod** (a crustacean) is the first intermediate host. **Fish** are the second intermediate hosts.	**Diphyllobothriasis** **Transmission** is by eating raw or undercooked **fish** containing larvae. Infection is usually asymptomatic, but may cause diarrhea. May cause a **deficiency of vitamin B$_{12}$** which results in megaloblastic anemia.

pathogen	appearance	life cycle	diseases
Echinococcus granulosus DOG TAPEWORM	**One of the smallest tapeworms**. Consist of a **scolex** and only **three proglottids**. **Diagnosis** relies on detection of serologic antibodies.	• **Adults** live and produce **eggs** in the intestines of *dogs*. Adults live for 6 months to 1 year and grow to about 5 mm. • *Sheep* or *cattle* become the intermediate hosts by ingesting contaminated dog feces. • The eggs become **embryos** (oncospheres) which migrate throughout the body where they develop into **cysts** (known as hydatids). • The cycle is completed when the dog ingests lamb or beef infected with cysts. **Dogs** are the definitive hosts and **sheep** are the intermediate hosts. Humans are dead end hosts. **Liver** is the most common location of hydatid cysts.	**Hydatid cyst disease** **Transmission** occurs when humans come in contact with dog feces contaminated with eggs. The ingested eggs become embryos (oncospheres) in the intestine. The embryos migrate to various organs throughout the body (i.e. liver, lung, brain) where they develop into large **cysts**, known as **hydatids**. Cysts enlarge at a rate of 1 cm/year. The gradually enlarging masses compress adjacent tissue and cause symptoms: liver failure, hemoptysis (coughing-up blood), and headaches. Rupture of the cysts may cause a fatal **anaphylaxis** or rapid spread of organisms. **Treatment** involves surgical removal of the cysts.

Helminth, Platyhelminth, Cestodes (tapeworms)   The Microbiology Companion, Topf and Faubel ©1997   **199**

Parasites

Parasites

Trematodes also known as **flukes** are a type of platyhelminth (flatworm).

Platyhelminthes (flatworms)

Cestodes (tapeworms)

Taenia solium

Taenia saginata

Diphyllobothrium latum

Echinococcus granulosus

Trematodes (flukes)

Schistosoma (blood fluke)

mansoni

japonicum

hematobium

Clonorchis sinensis (liver fluke)

Paragonimus westermani (lung fluke)

Trematodes (flukes) are flat, bilaterally symmetric worms which range from a few millimeters to several centimeters in length. They have two suckers. One is the mouth and the other is used for attachment and locomotion. The trematodes do not have an anus and simply regurgitate undigested food. The trematodes are divided into two groups based on their sexual organs: the schistosomes (the genus *Schistosoma*) and the hermaphrodites (the genus *Clonorchis* and *Paragonimus*). The schistosomes exist as two sexes and produce nonoperculated eggs. The male and female schistosomes live attached to each other. Hermaphrodites contain both male and female organs and produce operculated eggs.

The life cycle of the trematodes has human and snail hosts. **Humans are the definitive hosts** (maturation from larvae to adult) and **snails are the intermediate hosts** (maturation from eggs to larvae). The hermaphrodite trematodes also have a second intermediate host as part of their life cycle (crabs for *Paragonimus* and fish for *Clonorchis*). Transmission to humans is by ingestion of larvae in undercooked fish or crab (*Clonorchis* and *Paragonimus*) or by penetration through the skin with cercariae *Schistosomes*).

Life cycle summary:
◆ Maturation from eggs to larvae occurs in the intermediate host (snails).
◆ Hermaphrodites have a second intermediate host.
◆ Maturation from larvae to adult occurs in the definitive host (people).

Helminth, Platyhelminth, Trematodes (fluke)   The Microbiology Companion, Topf and Faubel ©1997   **202**

Summary of Trematodes (flukes)

Trematode	eggs	life cycle	disease
Schistosoma		Humans and snails.	**Schistosomiasis** **Swimmers itch** Common in fresh water lakes of U.S. Occurs when humans are infected with cercariae whose proper host is some other animal. Itchy lesion at site of entry.
Schistosoma mansoni	prominent lateral spine on eggs	**Humans** and **snails**. GI tract: adults live in the inferior mesenteric veins from the descending colon and rectum.	**Schistosomiasis**. Rash, fever and chills followed by GI hemorrhage and hepatosplenomegaly.
Schistosoma japonicum	very small lateral spine on eggs	**Humans** and **snails**. GI tract: adults live in the superior mesenteric veins from the small intestine and ascending colon.	**Schistosomiasis**. Rash, fever and chills followed by GI hemorrhage and hepatosplenomegaly.
Schistosoma hematobium	terminal spine on eggs	**Humans** and **snails**. Urinary tract: adults live in the inferior mesenteric veins.	**Schistosomiasis**. Rash, fever and chills followed by hematuria, and possibly bladder cancer.
Clonorchis sinensis	eggs are small, brown and operculated	**Humans** are the definitive host. **Snails** are the first intermediate host. **Fish** are the second intermediate host.	**Clonorchiasis** (Oriental liver fluke infection): fluke travels through intestinal wall and grow into adults in the biliary ducts.
Paragonimus westermani	eggs are operculated	**Humans** are the definitive host. **Snails** are the first intermediate host. **Crabs** are the second intermediate host.	**Paragonimiasis**: fluke travels through the intestinal wall to settle in the lung. Symptoms include cough with bloody sputum.

All trematodes may be successfully treated with praziquantel.

pathogen	appearance	life cycle	disease
Schistosoma	There are both female and male schistosomes. The female lives in a groove on the male's body so he can fertilize her eggs. **Sexual reproduction** occurs in a human host. **Asexual reproduction** occurs in freshwater snails and results in the production of cercariae. Cercariae are the infectious form and have forked tails.	**Transmission** to humans occurs when the **cercariae** form of *Schistosoma* penetrates through human skin. In the skin, the cercariae differentiate into **larva**. The larva travel to the *portal veins* and become adults and mate in the liver. After mating, the different species migrate through the veins of the GI tract (*mansoni* and *japonicum*) or urinary tract (*hematobium*). Upon reaching their destination, the female lays eggs (that are immediately fertilized by the attached male), which penetrate the endothelial walls to enter the gut or bladder. Eggs are excreted in urine or feces. When the eggs reach fresh water, they hatch and infect snails. The organism replicates in the snail and produces cercariae which are able to infect humans. The organism is widespread, over 200 million persons are infected worldwide.	**Schistosomiasis** Infection is characterized by an intensely itchy rash. Disappearance of the rash 1-3 weeks later is followed by fever, headache and diarrhea. One to two months after the initial infection, the eggs induce an immune response which resembles serum sickness. Symptoms include: fever, chills, diarrhea, hepatomegaly and splenomegaly. Eggs reaching the vasculature of the intestines or bladder induce inflammation, granuloma formation, fibrosis and scarring. The tissue response may occlude blood vessels. May cause **portal hypertension**. **Swimmers itch** Is common in the fresh water lakes of the U.S. When humans are infected with cercariae whose proper host is some other animal the parasite dies and leaves an itchy lesion. The parasite is able to evade host defenses by covering itself with host antigens.

Platyhelminth, Trematode (fluke), *Schistosoma*   The Microbiology Companion, Topf and Faubel ©1997   **203**

Parasites

Platyhelminth, Trematode (fluke), *Schistosoma*   The Microbiology Companion, Topf and Faubel ©1997   **204**

pathogen	appearance	life cycle	diseases
Schistosoma mansoni	A **prominent lateral spine** is present on the eggs. Found in Africa and the Caribbean.	Affects the GI tract. Adult organisms travel through the **inferior** mesenteric veins to the **descending colon** and **rectum**.	Continued inflammatory response may cause GI hemorrhage and massive **hepatosplenomegaly**.
Schistosoma japonicum	A **very small lateral spine** is present on the eggs. Found in Japan and China.	Affects the GI tract. Adults travel through the **superior mesenteric veins** to the vessels of the small intestine and ascending colon.	Continued inflammatory response may cause GI hemorrhage and massive **hepatosplenomegaly**.
Schistosoma hematobium	A **terminal spine** is present on the eggs. Found in Africa and the Middle East.	Affects the urinary tract. Adults travel through the **inferior mesenteric** veins to the veins of the pelvic organs, particularly the **bladder**.	Continued inflammatory response may cause fibrosis and granulomas in the bladder wall. Symptoms include **hematuria** (blood in the urine) and **dysuria** (pain during urination). Secondary bacterial infections are common. Infection is associated with high rates of **bladder cancer**.

Think: **hema**tobium may cause **hema**turia.

pathogen	appearance	life cycle	diseases
Clonorchis sinensis LIVER FLUKE	**eggs**: small, brown and operculated **Diagnosis** is by finding eggs in stool. hermaphrodite	• The **larval** form (cercariae) are ingested by *humans* eating infected fish. • Maturation into the **hermaphrodite adult** occurs in the biliary ducts where **eggs** are produced. • Eggs from feces reach fresh water and are ingested by *snails*. • In the snail, the eggs mature into free swimming **cercariae** • The cercariae are released and lodge under the scales of *fish*. **Humans** are the definitive host. **Snails** are the first intermediate host. **Fish** are the second intermediate host.	**Clonorchiasis** (Oriental liver fluke infection) Common in China, Japan, Korea and Indochina where it affects about 20 million people. **Transmission** occurs by the ingestion of raw or undercooked **fish** contaminated with larvae. After ingestion and excystation, the fluke travels through the intestinal wall and settles into the **biliary ducts** where they grow into adults. Infection is usually asymptomatic. **Symptoms** in persons with a large worm load may include abdominal pain, weight loss and/or enlargement of the liver.

pathogen	appearance	life cycle	diseases
Paragonimus westermani LUNG FLUKE	**eggs:** operculated **Diagnosis** relies on finding eggs in sputum or feces. hermaphrodite	• **Larval** form ingested by *humans*. • After ingestion and excystation, the fluke travels through the intestinal wall and diaphragm to settle in the **lung**. • Hermaphroditic adults produce **eggs** in the lung. • Eggs are coughed-up and either expectorated or swallowed. • Eggs from sputum or feces reach fresh water and hatch into **miracidia.** • **Miracidia** are ingested by *snails* and become **cercariae** which infect *crabs*. • Crabs are eaten by humans and the cycle continues. **Humans** are the definitive host. **Snails** are the first intermediate host. **Crabs** are the second intermediate host.	**Paragonimiasis** **Transmission** occurs by the ingestion of raw or undercooked crab meat contaminated with larvae. The fluke travels through the intestinal wall and the diaphragm to settle in the **lung**. **Symptoms** include cough with bloody sputum.

Nematodes are roundworms. They do not have a segmented body. Their multicellular bodies are protected by a tough coating called a **cuticle.** The adult form has two sexes which only exist and mate when within a host. The female produces eggs which become larvae within its uterus or outside of the host. The larvae then use a variety of strategies to infect hosts. The nematodes are categorized by the primary site of infection: **intestines** or **tissues**.

Intestinal nematodes	Tissue nematodes
Enterobius vermicularis	Trichinella spiralis
Trichuris trichiura	Ancylostoma braziliense
Ascaris lumbricoides	Ancylostoma caninum
Hookworms	Toxocara canis
Ancylostoma duodenale	Wuchereria bancrofti
Necator americanus	Brugia malayi
Strongyloides stercoralis	Dracunculus medinensis
	Onchocerca volvulus
	Loa loa

Intestinal nematodes typically do not cause heavy parasitic load because the life cycle requires a maturation step outside of the host. This means the only way for a worm to infect a person is for the person to ingest eggs. Because it is impossible for the parasite to multiply entirely within the host, high parasite loads can only result from repeated inoculation or a single massive ingestion of eggs.

Intestinal nematodes which live in the small intestine can cause malnutrition because they compete with the host for food. Nematodes that live in the colon cannot cause malnutrition because the host has already completed nutritional absorption. The presence of nematodes in the colon may cause diarrhea.

All intestinal nematodes may be successfully treated with albendazole.

Summary of Intestinal Nematodes

organism	appearance/life cycle	disease	important features

Enterobius vermicularis
 	
PINWORM
		Pinworm infection
		  hand to mouth transmission
		  perianal pruritus
			Most common helminth infection in the U.S.
			Humans are the only host

Trichuris trichiura
 	
WHIPWORM
	Eggs are barrel shaped with plugs.	**Whipworm infection**
		  fecal-oral transmission
		  usually asymptomatic	

Ascaris lumbricoides	Life cycle includes pulmonary migration of larvae.
	Adults do not attach to the mucosa of the intestines (all the other nematodes on this page do).
		Ascariasis
		  fecal-oral transmission
		  lung infection (larvae)
		  Loeffler's syndrome (larvae)
		  intestinal infection (adults)
			Most common helminth infection in the world.
			Largest nematode that infects humans.

Hookworms
  Ancylostoma duodenale
  Necator americanus
	Life cycle
	• filariform larvae penetrate skin
	• travel to pulmonary circulation
	• enter the alveoli, coughed-up and swallowed.
	• attach to intestines, feed on blood
		Hookworm infection
		  transmission by contact with skin
		  ground itch
		  pneumonitis
		  microcytic anemia
			The adult attaches to the villi of the small intestine and sucks blood.
			The worm produces an anticoagulant

Strongyloides stercoralis
 	
THREADWORM
	Life cycle (hookworm type, also has autoinfection type and free-living type)
	• filariform larvae penetrate skin
	• travel to pulmonary circulation
	• enter the alveoli, coughed-up and swallowed.
	• attach to intestines, feed on blood
		Strongyloidiasis
		  cutaneous lesions, ground itch
		  pneumonitis
		  intestinal symptoms
			Smallest intestinal nematode.
			The only intestinal nematode that can reproduce entirely within the host.

pathogen	appearance	life cycle	disease
Enterobius vermicularis PINWORM	**Females** are 8-13 mm long and cream colored with a pointed tail (pinworm). **Males** are 2-5 mm long and cream colored with a curved tail. **Diagnosis** is by recovery of eggs from the anal mucosa with **Scotch Tape**™.	Male and female **adults** live attached to the mucosa of the *cecum* where they feed, grow and copulate. When the female is ready to lay its **eggs**, it travels down the colon and deposits them **outside** of the *anus*. This usually occurs at night. The **eggs** are <u>sticky</u> and will attach to fingers or bed sheets. Because the eggs are <u>resistant to dehydration</u>, they can remain viable for days. After ingestion, the **eggs** hatch in the *upper small intestine* and the **larvae** travel to the cecum where they mature into adults. Life cycle is two weeks. **Humans** are the only known host.	**Pinworm infection** **Most common** helminth infection in the **U.S.** and western **Europe**. **Transmission** is **hand to mouth**. Ingestion of the eggs may occur after direct contact with perianal skin (itching causes autoinfection) or after handling clothing or bedding. Household members commonly become infected and infection is difficult to control, even with scrupulous hygiene. Transmission is not technically fecal-oral because the <u>eggs are not contained within the stool</u>. Infection is relatively benign. The most common symptom is **perianal itching**. Most commonly occurs in children 5-10 years old. Infection has been linked to vaginitis and UTIs in young girls. **Treatment** is reserved for symptomatic individuals.

Think: **p**inworm causes **p**erianal **p**ruritus.

Helminth, Intestinal Nematodes, *Enterobius*   The Microbiology Companion, Topf and Faubel ©1997   **209**

Parasites

pathogen	appearance	life cycle	disease
Trichuris trichiura WHIPWORM	The parasite is 3-5 cm long. The anterior 2/3 is thin and the last third is thicker in caliber, like a whip. The male has a curved tail. **Diagnosis** is very difficult in asymptomatic individuals but in symptomatic patients the **eggs are visible in the stool** and **eosinophilia occurs** in the blood. **Eggs** are very characteristic: barrel shaped with plugs at either end.	**Adults** attach to the wall of the *colon* with the thin end of the whip. The female releases between 3,000 and 10,000 **eggs** into the *colon* <u>per day</u>. The eggs are deposited with the *stool* and become infectious after 10 days. The **eggs** hatch after ingestion. The **larvae** then mature in the d*uodenum* for up to one month before traveling to the *cecum* and becoming adult worms. The organism is most common in **warm, moist climates** which allow the eggs to survive outside a host for a long period of time. Proper disposal of feces is the best prevention.	**Whipworm infection** **Transmission** is by ingestion of eggs contained in soil which has been contaminated by human feces (<u>fecal-oral</u> transmission). Infection is a problem in areas without sewers or where human feces is used as fertilizer. Infection in the U.S. occurs in the **southeastern states** where about 2 million people are infected. The infection is found primarily among **toddlers**. Worldwide it is estimated that about 500 million people are infected. Infection is usually with a <u>low worm load</u> and is <u>asymptomatic</u>. Repeated infection may result in a <u>large worm load</u> and cause <u>anemia</u>, <u>abdominal pain</u> and <u>diarrhea</u>. **Treatment** is reserved for symptomatic individuals. Treatment rarely cures a person but reduces the parasite load to render the patient asymptomatic.

pathogen	appearance	life cycle	disease
Ascaris lumbricoides	The worm is the size and shape of an earthworm. It typically measures 15 to 35 cm in length. **Diagnosis** is based on finding eggs in feces or larvae and eosinophils in the sputum. *A. lumbricoides* is the largest nematode that infects humans.	**Adults** live in the *small intestine.* where they remain free and <u>do not attach to intestina mucosa</u>. The female releases 200,000 **eggs** <u>per day</u>. The eggs are passed with the *feces*. The **eggs** live in the *soil* for three weeks before becoming infectious. They are then ingested and mature into **larvae**. The **larvae** penetrate the intestinal walls to enter the portal circulation and systemic veins. They are carried through the right side of the heart and into the *pulmonary circulation*. In the *pulmonary circulation* the larvae are too big to pass through the capillaries and migrate into the *alveolar air spaces*. The larvae are subsequently <u>coughed up and swallowed</u>. Then, in the *intestine*, they mature into the **adult** form. The organism is most common in **warm climates** which allow the eggs to survive outside a host for a long period of time. Proper disposal of feces is the best prevention.	**Ascariasis** **Most common intestinal helminth infection in the world.** Worldwide, over 1 billion people are infected. **Transmission** is by ingestion of eggs contained in fecally contaminated soil. Disease results from **larvae** in the **lung** and <u>adults</u> in the <u>intestines</u>. **Lungs** (infection by larvae) **Larval migration** from the pulmonary circulation into the pulmonary air spaces causes bleeding and the accumulation of **exudate** in the alveoli. The resulting symptoms include fever, cough, wheezing and hypoxia. **Loeffler's syndrome** is a pulmonary condition characterized by fever and a productive cough. It occurs in previously infected individuals and is thought to be the result of a **hypersensitivity** reaction to the parasitic antigens. **Intestines** (infection by adults) Infection with a <u>low worm load</u> is <u>asymptomatic</u>. During episodes of **fever**, however, the worms become motile and may cause symptoms 2° to **obstruction** of the bile duct, pancreatic duct or appendix. Patients with frequent reinfection and a large amount of worms may have abdominal symptoms such as pain, malabsorption and/or obstruction.

Think: *A. lumbricoides* causes **l**ung disease in the **l**arval stage, repeated infections may result in **L**oeffler's syndrome.
   A. lumbricoides is the <u>largest</u> nematode and infects the <u>largest</u> number of people.

pathogen	appearance	life cycle	disease
Hookworms Two types: *Ancylostoma duodenale* OLD WORLD HOOKWORM *Necator americanus* NEW WORLD HOOKWORM	Hookworms are 1 cm long and shaped like an "S." Diagnosis is made by: • observing eggs in the stool • eosinophilia • occult blood in the stool *Ancylostoma* has **four teeth** which it uses to attach to the intestinal wall. *Necator* has **knife-like fins** called **cutting plates** which it uses to attach to the intestinal wall.	**Adults** attach to the *small intestine* and feed on blood. The worm produces an <u>anticoagulant</u> which causes additional blood loss. The female releases 10,000 **eggs** <u>per day</u> which are passed in the *stool*. The eggs hatch in the *soil* releasing **rhabditiform larvae** which survive in the soil by eating bacteria. After doubling in size, the larvae changes form to become the infectious **filariform larvae**. The **filariform larvae** penetrate the *skin* of a human and enter the lymphatic system. They travel with the lymph to the superior vena cava and then into the *pulmonary circulation*. The larvae break into the *alveoli* and are <u>coughed up and swallowed</u>. In the small intestine, the larvae attach to the *villi* and mature into the **adult** form. The worm feeds on blood.	**Hookworm infection** Worldwide, approximately 900 million people are infected. **Transmission** occurs when skin is exposed to the infectious **filariform larvae** (i.e. when walking barefoot on the soil). Infection is usually asymptomatic. Symptoms that do occur depend on the site of the organism: <u>Skin</u> "**Ground itch**" is an intensely pruritic **skin** rash characterized by edema and erythema which occurs at the site of entry. <u>Lung</u> **Pneumonitis** may occur during **pulmonary** migration of larvae. <u>Intestines</u> **Microcytic anemia** and hypoalbuminemia may occur with large worm burdens secondary to chronic blood loss from the mucosa of the **intestines**. Supplemental **iron** is used to treat the anemia.

pathogen	appearance	life cycle	disease
Strongyloides stercoralis THREADWORM	The worm is the **smallest intestinal nematode** measuring only 3 mm. **Diagnosis** is by observation of the larvae in stool or sputum. The larvae are intermittently released into the stool requiring multiple stool samples to find the parasite.	The life cycle of *Strongyloides stercoralis* is the most varied of all parasites. It can be broken down into three types: **Hookworm type** **Adults** live in the wall of the *duodenum*. The female releases **eggs** within the wall of the duodenum. The eggs hatch and **rhabditiform larvae** are passed in the *stool*. The larvae can exist in the soil by feeding on bacteria before transforming into infectious **filariform larvae**. The larvae penetrate the *skin* of a host, go through the lymphatics to the lungs, and get <u>coughed up and swallowed</u> just like the hookworm larvae. **Autoinfection type** The **rhabditiform larvae** are released from the duodenal wall just as in the hookworm type. The difference is that the feces are not passed immediately secondary to constipation. The rhabditiform larvae transform into the infectious **filariform larvae** <u>while still in the body</u>. The filariform larvae penetrate the bowel wall or perianal skin, travel to the lungs and are <u>coughed up and swallowed</u>. This is the **only** example of an **intestinal nematode reproducing without exiting the body**. Thus, massive parasite loads can develop without reinfection. This life cycle is common in people with depressed T-cell immunity (i.e. AIDS). **Free-living type** Occasionally the **rhabditiform larvae** can mature into an **adult** form <u>outside of a host</u>. The adult is able to propagate and survive for multiple generations without a host. The life cycle still includes the infectious **filariform larvae**. *Strongyloides* is most common in warm, damp climates.	**Strongyloidiasis** **Transmission** occurs when the infectious filariform larvae come in contact with **skin**. Infection is usually asymptomatic. Symptoms that do occur depend on the site of the organism: <u>Skin</u> **Skin lesions** may occur at the site of entry. <u>Lungs</u> **Pneumonitis** may occur during pulmonary migration of larvae. <u>Intestines</u> **Intestinal symptoms** may include diarrhea, malabsorption, and abdominal pain. In **immunocompromised** individuals with autoinfection, every organ is parasitized and secondary bacterial infection is common.

Tissue Nematodes induce disease by their presence in the tissues of the host. Three of these nematodes infect humans only by accident and can not complete their life cycle; humans are dead end hosts. These parasites, *Trichinella spiralis, Toxocara canis, Ancylostoma braziliense*, though unable to complete their life cycle still cause disease nonetheless. The remainder of the tissue nematodes have life cycles which use humans as hosts and cause disease depending on their location in the body. *Wuchereria bancrofti* and *Brugia malayi* live in the lymphatic system; *Dracunculus medinensis, Onchocerca volvulus* and *loa loa* live in the subcutaneous tissues.

Summary of Tissue Nematodes

organism	appearance location of lesion	disease	important features
Trichinella spiralis	Muscles, heart.	**Trichinoses** **Transmission** by eating pork infected with cysts.	Humans are dead-end hosts.
Ancylostoma braziliense *Ancylostoma caninum*	Subcutaneous tissues, pulmonary infiltrates in the lungs.	**Cutaneous larva migrans** **Transmission** by larvae contact with skin. Disease caused by larvae travelling in the subcutaneous tissues under the skin.	Humans are dead-end hosts.
Toxocara canis	Liver, lungs, heart and eyes.	**Visceral larva migrans** **Transmission** by ingestion of eggs. Pulmonary symptoms. Blindness.	Humans are dead-end hosts.
Wuchereria bancrofti	Threadlike appearance. Microfilaria are present in the blood only at night. Lymph vessels of legs and genitals.	**Bancroftian filariasis** Tender, swollen lymph nodes. **Obstructive filariasis** (elephantiasis) Edema, swelling of genitalia and legs.	Adults live in the lymphatic system. **Humans** are the only definitive hosts. **Mosquitoes** are the intermediate hosts and the method of **transmission** (vector).

organism	appearance location of lesion	disease	important features
Brugia malayi	Threadlike appearance Microfilariae have a sheath. Lymph vessels of legs and genitals.	**Malayan filariasis** Symptoms are similar to bancroftian filariasis.	**Humans** are the definitive hosts. **Mosquitoes** are the intermediate hosts.
Dracunculus medinensis	Female worms grow up to one meter. Subcutaneous tissue and skin.	**Dracunculiasis** Ulcers and blisters. Remove worms by wrapping them around a stick.	Adults live in the subcutaneous tissues of humans. **Humans** are the definitive hosts. **Copepods** are the intermediate hosts.
Onchocerca volvulus	Microfilariae do not have a sheath. Subcutaneous tissue and eyes.	**Onchocerciasis** Dermal nodules. **River blindness**	Microfilariae live in the subcutaneous tissues of humans. **Humans** are the only definitive hosts. **Black flies** are the intermediate hosts.
Loa loa	Microfilariae have a sheath. Subcutaneous tissue and eyes.	**Loiasis** Calabar swellings. Crawl across the eye.	Adults live in the subcutaneous tissues of humans. **Humans** and **monkeys** are the only definitive hosts. **Deer flies** are the intermediate hosts.

pathogen	appearance	life cycle	disease
Trichinella spiralis	The **male** is 1.5 mm long. The **female** is 3.5 mm long. **Diagnosis** is by detection of persistent eosinophilia (15-50% of leukocytes). ELISA can detect antibody formation within one week of infection.	**Adults** live and mate in the mucosa of the small intestines of *carnivores*. After mating, the male dies. The female releases **larvae** instead of eggs. The female produces approximately 1500 eggs (in her 1-3 month lifetime) which hatch into larvae while still in the uterus. The larvae penetrate the gut wall and enter the systemic circulation. Larvae enter *striated muscle* (skeletal and cardiac) and **encyst** themselves. The larvae remain viable for up to 10 years. **Transmission** occurs when the infected muscle is eaten by a predator. Since humans are rarely eaten, they are dead-end hosts. In humans, the **disease is caused by the larval form**, not the adult parasite.	**Trichinoses** **Incidence is 150,000 to 300,000 cases** annually with only about 100 symptomatic cases a year. **Transmission** occurs by eating undercooked infected meat, usually **pork**. Other sources include home made sausage, bear meat and pig roasts (in Hawaii). Symptoms include nausea, abdominal pain and diarrhea 1-2 days after ingestion. After one week, the larvae begin to enter the muscles and the patient has fever, **muscle pain** and weakness. Infection of the heart may cause EKG abnormalities, increased heart rate and **CHF**. **Respiratory failure** may occur secondary to muscle damage. **Treatment** with mebendazole will destroy tissue larvae but can trigger a hypersensitivity reaction to the larval antigens. This can be prevented with prophylactic administration of steroids.

pathogen	life cycle	disease
Ancylostoma braziliense CAT HOOKWORM *Ancylostoma caninum* DOG HOOKWORM	**Adults** live in the *intestines* of **cats** and **dogs**. After **larvae** are deposited within the *stool*, they develop into **filariform larvae** which are capable of infecting humans. Filariform larvae penetrate the *skin* and travel in the *subcutaneous tissues*. The worms are incapable of completing their life cycle within humans. Disease is caused only by the **larval form**.	**Cutaneous larva migrans** **Transmission** occurs when the infectious filariform larvae comes in contact with **skin**. Clinical manifestations include an intensely pruritic, red, raised rash in a linear pattern 10 to 15 cm in length. 50% of patients get **Loeffler's syndrome**: eosinophilia with transient migratory pulmonary infiltrates. Patients can get secondary bacterial infections from scratching too much. **Treat** with albendazole.
Toxocara canis	**Adults** live in the *intestines* of **dogs**, **foxes** and **wolves**. The adult female releases 200,000 **eggs** a day which become infectious after 2-3 weeks in the *soil*. The **eggs** are ingested by *canines*. In the canine, the larvae enter the alveolar sacs, and are <u>coughed up and swallowed</u>. In *humans*, the larvae cannot enter the alveolar spaces. Instead the larvae mature in the *systemic circulation*. As they get larger, they enter various tissues and cause necrosis. The parasite is unable to enter the intestines to complete its life cycle. Disease is caused by the **larval form**. **Humans** are **dead-end hosts**.	**Visceral larva migrans** **Transmission** occurs when eggs in the soil are ingested. The parasite most often enters the liver, lungs, heart and skeletal muscle. It induces necrosis, bleeding and granuloma formation. **Pulmonary symptoms** such as coughing and wheezing are early signs of infection. **Blindness** may result secondary to retinal involvement if the larvae enter the eye. The disease is common in the U.S. with 4-20% of the population having serologic evidence of past exposure. More common in the Southeastern U.S. Diagnosis is by liver biopsy. ELISA tests are available. Treat with diethycarbamazine.

pathogen	appearance	life cycle	disease
Wuchereria bancrofti	Threadlike appearance. The **female** is 100 mm long. The **male** is 40 mm long. The microfilariae have a sheath. **Diagnosis** is by observation of microfilariae in the blood. Microfilariae are at their highest concentration in the bloodstream between 10 p.m. and 2 a.m. which is when blood samples should be taken.	**Adult worms** infect the lymph nodes and vessels primarily in the lower extremities and genital area of *humans* where the male and female worms mate. The females release **microfilariae** (larval form). The life cycle is unusual because there is no egg stage in development. The **microfilariae** enter the *blood stream*. By day the microfilariae concentrate in the pulmonary circulation and at night they enter the systemic circulation. At night, the **microfilariae** are ingested by *mosquitoes* as they feed. Within the mosquito, the microfilariae mature into **infectious larvae** in ten days. **Transmission** occurs when the mosquito injects the **infectious larvae** into the skin. The larvae penetrate the skin and enter the *lymphatic system* where they reside for up to a year before emerging in the adult form. **Humans** are the definitive hosts. **Mosquitoes** are the intermediate hosts.	**Bancroftian filariasis** Most of the symptoms result from the presence of the adult worm in the lymphatics. **Transmission** occurs when larvae are injected into the skin by an infected mosquito. Usually asymptomatic. Acutely, some patients present with **lymphangitis** (infected lymph vessels, causes red streaks on the skin which are not raised but follow the course of the lymph vessels). Symptoms of light infections are generally limited to **enlarged and tender lymph nodes.** Larger parasite loads, due to repeat infection, lead to fever, vomiting, tender lymphadenopathy and draining ulcers. Most of these symptoms are a result of a **hypersensitivity** reaction against parasite antigens. **Elephantiasis** (obstructive filariasis) Occasionally after many repeat infections the immune response to the worm in the lymph tract causes **obstruction** of the lymph vessels. The obstruction of lymph flow results in massive **edema and swelling** which is usually located in the legs and genitalia (elephantiasis). **Treatment** with diethylcarbamazine kills the microfilariae but not the adult worms, making it an effective prophylaxis.

pathogen	appearance	life cycle	disease
Dracunculus medinensis GUINEA FIRE WORM	The **female** worms grow up to <u>one meter</u> in length. **Diagnosis** is by seeing the worm under the skin.	**Adults** live in the *subcutaneous tissues* of humans. After mating, a portion of the female's body forms an **ulcer** in the skin which facilitates the release **larvae** when <u>underwater</u>. The **larvae** are ingested by *copepods* (crustaceans) in water. The larvae mature into the **infectious form** within this intermediate host. **Transmission** occurs when copepods are swallowed by *humans* drinking contaminated water. The larvae penetrate the gut mucosa to enter the *lymphatics and subcutaneous tissues*. The larvae develop into adults over a year. **Humans** are the definitive hosts. **Copepods** are the intermediate hosts.	**Dracunculiasis** Disease is characterized by **skin** manifestations which include **ulcers and blisters** which may be pruritic. Hypersensitivity reactions to the parasitic antigens are common and result in **systemic** symptoms such as vomiting, diarrhea and respiratory distress. **Treatment** is by physical removal of the adult worm from the subcutaneous tissues. A **stick** is inserted in the ulcer and turned so that the worm wraps around it. Care must be taken not to break the worm, therefore only a few centimeters are wrapped around the stick a day.

pathogen	appearance	life cycle	disease
Onchocerca volvulus RIVER BLINDNESS	Adults are 2-50 mm. The microfilariae do not have a sheath. **Diagnosis** is by recognition of microfilariae in a skin biopsy or visualization of microfilariae in the eye by slit lamp.	**Adults** live in the *subcutaneous tissue* of humans in areas called **dermal nodules**. The female produces 2000 **microfilariae** daily. The **microfilariae** migrate through the *subcutaneous tissues* for up to two years. **Microfilariae** are taken up by the *black fly* during feeding. Within the fly, microfilariae become **infectious larvae**. **Transmission** to humans occurs when the *black fly* injects **infectious larvae** into the skin during feeding. Within a human, the larvae mature for 6- 12 months before becoming viable **adults**. Most common in **Africa**. **Humans** are the only definitive hosts. **Black flies** are the intermediate hosts.	**Onchocerciasis** Clinically the patient presents with multiple hard, mobile, non-tender **subcutaneous nodules**. They are usually 1-3 cm in diameter. Most symptoms are due to the **microfilarial** wanderers which induce a hypersensitivity reaction. This causes pruritus, a papular rash and **thickened skin**. "**River blindness**" is the most devastating complication which occurs when the microfilariae concentrate in the eyes. (It is called river blindness because black flies which transmit the disease breed near rivers.) **Treatment** is with ivermectin to kill the microfilariae.
Brugia malayi	Threadlike appearance. The microfilariae have a sheath. Morphologically distinct from *Wuchereria bancrofti*.	Life cycle is identical to *Wuchereria*. Humans and other mammals are the definitive hosts.	**Malayan filariasis** The clinical picture is the same as for bancroftian filariasis. Treatment is the same.

Parasites

pathogen	appearance	life cycle	disease
Loa loa AFRICAN EYE WORM	The microfilariae have a sheath. **Diagnosis** is by observation of microfilariae in the blood.	**Transmission** to *humans* occurs when **infectious larvae** are injected into the skin by the bite of a *deer fly*. The infectious larvae mature into **adults** in the *subcutaneous tissues*. The adult lives in the subcutaneous tissues all over the body. As the **adults** migrate within the *subcutaneous tissue*, they release sheathed **microfilariae** which are taken up by the bloodstream. The **microfilariae** are ingested by a *deer fly* when feeding on an infected human. The **microfilariae** develop into **infectious larvae** within the fly. **Humans** and **monkeys** are the only definitive hosts. **Deer flies** are the intermediate hosts.	**Loiasis** **Calabar swellings** appear in the skin and are the result of inflammation caused by the migration of the adult worm in the **subcutaneous tissues**. The adult worm may **migrate across the surface of the eye** causing pain, tearing and anxiety. This highly disconcerting event does not result in permanent eye damage. Treat with diethylcarbamazine.

Think: a loa loa, oh baby, I'm a tissue nematode. Yeah, yeah, yeah, yeah, yeah.
　　　　a loa loa, oh baby, I'm a worm that's really bogue. Yeah, yeah, yeah, yeah, yeah.
　　　　a loa loa, oh baby, I can crawl across your eye. Yeah, yeah, yeah, yeah, yeah.
　　　　a loa loa, oh baby, Don't get bit by a deer fly. Yeah, yeah, yeah, yeah, yeah.

The fun loving gals and guys at Alert and Oriented Publishing are *still* offering a $25 prize (were fun, not rich) and immortality in the next edition of the book, to the best rendition of the Loa Loa song. Send you demo tapes to:
　　　　　　Alert and Oriented Publishing
　　　　　　C/O I do have a life outside of medical school contest
　　　　　　616 Church Street #3
　　　　　　Ann Arbor, MI 48104

Section five: Cross-Reference

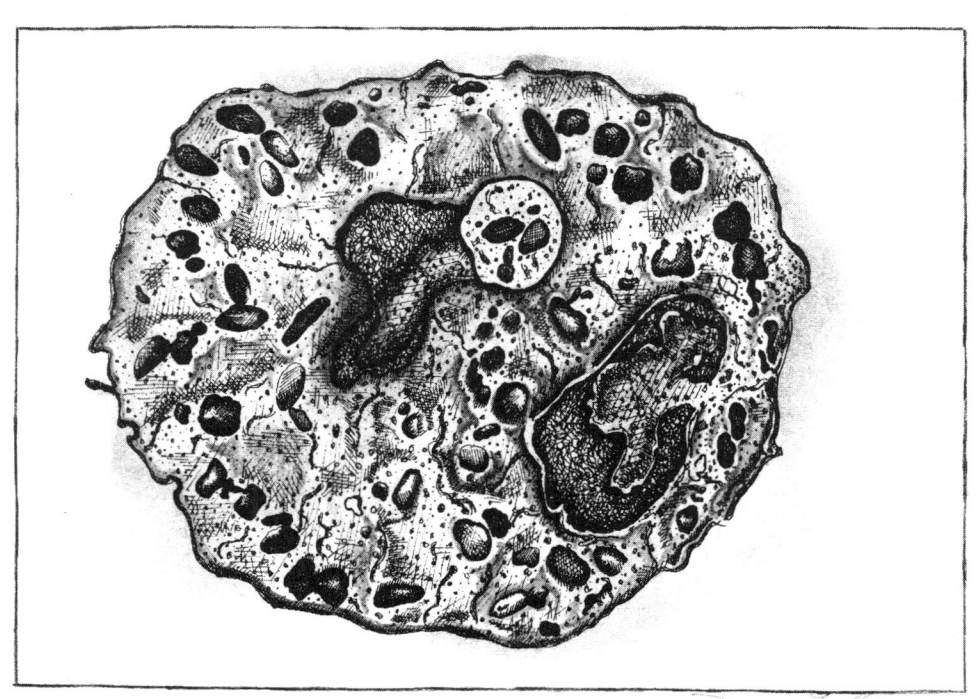

PMN, Dennis Franklin

Cross-Reference

Characteristics

Infections Which Cause Granulomas..............................226
Pathogens With a Capsule 226
Intracellular Reproduction and Survival......................226
Motile...226
Hemolysis (beta or alpha hemolytic)............................227
Bacteria which are normal flora 227
Catalase positive organisms..228
IgA protease 228
Recurrent fevers..228
Fungi which are dimorphic...228
Fungi which are non-dimorphic.......................................228
Fungi which produce asexual spores (conidia) 228

Culture

Cannot be grown on artificial media or cell culture.......230
Grown only in cell cultures 230
Culture is dangerous to lab workers..............................230
Special culture requirements..230

GI disease

Toxins which cause disease outside of the GI tract........231
GI disease without a toxin 232
Diarrhea 233

Transmission

Pathogens with arthropod vectors 234
Infections acquired in utero 235
Infections acquired in the birth canal 235

Meningitis

Most common cause of bacterial meningitis 236
Most common cause of meningitis in special cases.......237
Interpreting spinal taps for meningitis 237

Immunology

IgG..241
IgM...241
IgA..242
IgD..242
IgE..242
hypersensitivity..243

Vaccines

passive immunizations 246
routine childhood vaccines 247
vaccination schedule 248
active vaccines 249

Infections which induce the formation of **granulomas**

Gram + rods
  Listeria monocytogenes

Gram - rods
  Francisella tularensis
  Brucella abortus, melitensis,
  suis

Spirochetes
  Treponema pallidum
  (gummas of tertiary syphilis)

Mycobacterium
  tuberculosis, leprae,
  marinum, scrofulaceum

Fungi
  Coccidioides immitis
  Histoplasma capsulatum
  Blastomyces dermatitidis
  Aspergillus fumigatus

Parasites
  Leishmania donovani,
    tropica, mexicana,
    braziliensis
  Schistosoma mansoni,
    japonicum, hematobium

Pathogens which can reproduce and/or survive **intracellularly**

Obligate intracellular parasites
  Chlamydia (all)
  Rickettsiae (all)
  Viruses (all)

Gram + rods
  Listeria monocytogenes

Gram - rods
  Brucella (all)
  Salmonella (all)
  Legionella pneumophila
  Yersinia pestis
  Francisella tularensis

Mycobacteria
  Mycobacteria tuberculosis,
    leprae

Molds
  Histoplasmosis capsulatum

Parasites
  Leishmania donovani
  Plasmodium falciparum,
    vivax, ovale, malariae

Bacteria with a **capsule**

All capsules are made out of polysaccharide except Bacillus anthracis which has a capsule made out of D-glutamate (protein).

Gram + cocci
  Streptococcus pneumoniae

Gram - cocci
  Neisseria meningitidis

Gram + rods
  Bacillus anthracis
    (D-glutamate)

Gram - rods
  Bacteroides (all)
  E coli
  Salmonella (all)
  Klebsiella pneumoniae
  Haemophilus influenza
  Bordetella pertussis
  Yersinia pestis

Fungi
  Cryptococcus neoformans

Bacteria which are **motile**

Gram - rods
  E coli
  Salmonella (all)
  Proteus, Providencia,
    Morganella
  Vibrio cholera
  Yersinia enterocolitica
  Pseudomonas aeruginosa

All Spirochetes
  Borrelia burgdorferi,
    recurrentis
  Leptospira interrogans
  Treponema pallidum

Gram + rods
  Listeria monocytogenes
  Bacillus cereus

Bacteria which produce **hemolysis** on a blood agar plate

Beta hemolysis represents complete lysis of erythrocytes and causes a **clear ring** around the colonies on a BAP.

Alpha hemolysis represents only partial lysis of erythrocytes and causes a **green ring** around the colonies on a BAP.

Gram + cocci

Staphylococcus aureus	beta hemolysis
Streptococcus pyogenes (Group A Strep)	beta hemolysis
Streptococcus agalactia (group B Strep)	beta hemolysis
Streptococcus pneumoniae	alpha hemolysis
Streptococcus Viridans group	alpha hemolysis
Enterococcus faecalis, faecium	alpha hemolysis (usually)
Streptococcus bovis	alpha hemolysis (usually)

Others

Listeria monocytogenes	beta hemolysis
Mycoplasma pneumoniae	beta hemolysis

Bacteria which are **normal flora**

Gram positive cocci

Staphylococcus epidermidis	skin
Streptococcus. agalactiae	vagina
Streptococcus faecalis	intestine
Streptococcus Viridans group	oropharynx

Gram positive rods

Clostridium perfringens	colon and vagina
Clostridium difficile	gut flora of 3% of the population
Corynebacterium diphtheriae	skin
Actinomyces israelii	oral cavity and intestines

Gram negative rods

E. coli	intestines
Klebsiella pneumonia	intestines
Enterobacter cloacae	intestines
Serratia marcescens	intestines
Proteus	intestines
Morganella	intestines
Providencia	intestines
Bacteroides fragilis	intestines and vagina
Bacteroides melaninogenicus	mouth, vagina and large colon
Moraxella catarrhalis	oral cavity
Fusobacterium	mouth, intestines and vagina

Fungus

Candida albicans	oral cavity, vagina

X-Reference

Catalase positive organisms

Staphylococcus	gram + cocci
Listeria	gram + rod
Serratia	gram- rod
Nocardia	acid fast anaerobe
Aspergillus	gram + aerobe

Bacteria which produce IgA protease

Streptococcus pneumoniae	gram + cocci
Neisseria meningitidis, gonorrhoeae	gram - cocci
Haemophilus influenza	gram - rod

Diseases with recurrent fevers

Brucella	undulating fever	zoonotic gram negative rod
Borrelia recurrentis	relapsing fever	loosely coiled spirochete
Plasmodium	malaria	sporozoan parasite
Trypanosoma brucei	sleeping sickness	hemoflagellate

Fungi which are dimorphic

Coccidioides immitis

Histoplasma capsulatum

Blastomyces dermatitidis

Paracoccidioides brasiliensis

Malassezia furfur

Cladosporium werneckii

Sporothrix schenckii

Fungi which are non-dimorphic

Candida albicans	exists only as a yeast
Cryptococcus neoformans	exists only as a yeast
Aspergillus fumigatus	exists only as a mold
Zygomycetes (Mucor, Rhizopus)	exists only as a mold

Fungi which produce asexual spores (conidia)

Coccidioides immitis	arthrospores
Blastomyces dermatitidis	
Paracoccidioides brasiliensis	
Candida albicans	chlamydospores
Aspergillus fumigatus	
Zygomycetes (Mucor, Rhizopus)	sporangiospores
Dermatophytes	
Sporothrix schenckii	
Histoplasma capsulatum	microconidia and tuberculate macroconidia

Notes

X–Reference

Culture

Cannot be grown on artificial media or cell culture	Grown only in cell cultures	Culture is dangerous to lab workers
Mycobacteria leprae can be grown in the armadillo or the mouse footpad. Human papilloma virus Hepatitis B virus	*Treponema pallidum* *Chlamydia pneumoniae, psittaci, trachomatis* *Rickettsiae rickettsii, akari, prowazekii, typhi, tsutsugamushi, coxiella burnetii*	*Francisella tularensis* *Coccidioides immitis*

Special culture requirements

Streptococcus	Blood agar plate is used because it demonstrates hemolysis.
Neisseriaceae	Thayer Martin medium inhibits the growth of normal flora and other bacteria by the addition of antibiotics to the media.
Corynebacterium diphtheriae	Loeflers or BAP with potassium tellurite.
Gram - rods (enterics)	MacConkeys agar inhibits the growth of gram positive organisms and distinguishes lactose and non-lactose fermenters (lactose fermenters appear red). EMB (eosin methylene blue agar).
Bacteroides, Fusobacterium	Require anaerobic growth.
Haemophilus	Chocolate agar with factor X (heme) and factor V (NAD).
Legionella pneumophila	Mueller Hinton culture medium must contain iron and cysteine.
Bordetella pertussis	Bordet-Gengou medium.
Brucella	Culture media must be enriched.
Mycobacterium	Löwenstein-Jensen medium is used to culture all species of Mycobacterium.

Toxins which cause disease outside of the GI tract

pathogen	disease	toxin
Bacillus anthracis	Anthrax	Complex toxin: edema factor, lethal factor and protective antigen; ↑ cAMP.
Bordetella pertussis	Whooping cough	Pertussis toxin: irreversible activation of G proteins which ↑ cAMP and phospholipase C activity.
Clostridium botulinum	Botulism	Botulinum toxin: blocks the release of acetylcholine from the neuromuscular junction; flaccid paralysis.
Clostridium tetani	Tetanus	Tetanus toxin: blocks the release of inhibitory neurotransmitter; spastic paralysis.
Corynebacterium diphtheriae	Diphtheria	Diphtheria toxin: inactivation of EF-2 by ADP ribosylation.
Pseudomonas aeruginosa	Swimmer's ear / Nosocomial infections	Exotoxin A: inactivation of EF-2 by ADP ribosylation.
Staphylococcus aureus	Toxic shock syndrome	TSST-1 toxin.
Streptococcus pyogenes	Scarlet fever	Erythrogenic toxin.
Yersinia pestis	The plague	Murine toxin: irreversible shock and death in rodents.

Toxins

X–Reference

The Microbiology Companion, Topf and Faubel ©1997

Toxins causing disease in the GI tract

pathogen	disease	toxin	
Bacillus cereus	Food poisoning short incubation: 1-6 hr long incubation: 10-12 hr	Heat stable toxin: vomiting. Heat labile toxin: ↑cAMP, diarrhea.	unable to isolate organism in stool
Campylobacter jejuni	Enterocolitis	Heat labile enterotoxin: ↑cAMP.	
Clostridium difficile	Pseudomembranous colitis	Toxin A: diarrhea. Toxin B: cytotoxic.	
Clostridium perfringens	Food poisoning 6-16 hr incubation	Enterotoxin: heat labile acts in small bowel. Contaminated meat and poultry.	36 hours of low volume diarrhea
E. coli enterotoxigenic	Traveler's diarrhea	Heat stable toxin: ↑cGMP. Heat labile toxin: ↑cAMP.	
E. coli enteroinvasive	Hemolytic colitis	Shiga-like toxin: kills intestinal cells by inactivation of the 60S ribosome.	
Shigella	Dysentery	Shiga toxin: kills intestinal cells by inactivation of the 60S ribosome.	
Staphylococcus aureus	Food poisoning short incubation: 1-8 hr	Enterotoxin: heat stable; vomiting is a prominent symptom.	unable to isolate organism in stool
Vibrio cholera	Cholera	Choleragen: ↑cAMP.	
Vibrio parahaemolyticus	Food poisoning (shellfish)	Toxin similar to choleragen: ↑cAMP.	
Yersinia enterocolitica	Enterocolitis	Heat stable toxin: ↑cGMP.	

GI disease without a toxin

E. coli enteropathogenic	Dysentery
E. coli enterohemorrhagic	Diarrhea
Salmonella	Dysentery

Diarrhea

pathogen	fecal leukocytes	type of stool
Clostridium difficile	PMNs	**Pseudomembrane colitis** diarrhea with pus and mucus.
E. coli invasive (EIEC)	PMNs	**Dysentery**: Small volumes of diarrhea with pus and mucus which occurs after an incubation of 1-4 days.
E. coli toxigenic (ETEC)	none	Acute onset of voluminous **watery diarrhea** without mucus or blood.
E. coli pathogenic (EPEC)	PMNs massive	**Dysentery**: Small volumes of diarrhea with pus and mucus which occurs after an incubation of 1-4 days.
E. coli hemorrhagic (EHEC)	RBCs	Acute onset of voluminous **watery diarrhea** with blood.
Shigella	PMNs massive	**Dysentery**: Small volumes of diarrhea with pus and mucus which occurs after an incubation of 1-4 days.
Salmonella typhi	monocytes	**Enteric fever**: gradual onset of abdominal pain, fever and mild diarrhea lasting about one month. Usually associated with bacteremia.
Salmonella (other than *typhi*)	PMNs	**Dysentery**: Small volumes of diarrhea with pus and mucus which occurs after an incubation of 1-4 days.
Vibrio cholera	none	**Rice water stools**: copious **watery diarrhea** with flakes of mucus.
Campylobacter fecalis	PMNs	**Dysentery**: small volumes of malodorous bloody diarrhea with pus and mucus. This is associated with severe abdominal pain that can mimic appendicitis.
Yersinia enterocolitica	PMNs and monocytes	**Enteric fever**: gradual onset of abdominal pain, fever and mild diarrhea lasting about one month. Usually associated with bacteremia.
Viral diarrhea	none	Acute onset of voluminous **watery diarrhea** with blood. Often with fever, nausea and vomiting.
Giardia lamblia	none	**Watery diarrhea** which is voluminous and non-bloody. There are cramps, malabsorption and flatulence. The course can last for months.
Entamoeba histolytica	both in low numbers	**Dysentery**: small volumes of diarrhea with pus and mucus which occurs after an incubation of 1-4 days.

Pathogens with **arthropod vectors**

Gram - rods	disease	vector	reservoir
Yersinia pestis	Bubonic plague	Fleas	Rodents
Francisella tularensis	Tularemia	Ticks	Wild animals

Spirochetes

Borrelia burgdorferi	Lyme disease	Deer ticks	Deer
Borrelia recurrentis	Relapsing fever	Lice	Humans

Rickettsiae

Rickettsiae rickettsii	Rocky Mountain spotted fever	Ticks	Ticks
Rickettsiae akari	Rickettsialpox	Mites	Mice
Rickettsiae prowazekii	Typhus fever	Lice	Humans
Rickettsiae typhi	Typhus fever	Fleas	Rodents
Rickettsiae tsutsugamushi	Scrub typhus	Mites	Rodents

Rickettsiae *coxiella burnetii* is the only Rickettsial infection not transmitted by an arthropod vector.

Viruses

Eastern equine encephalitis virus	Eastern equine encephalitis	Mosquitoes	Wild birds
Western equine encephalitis virus	Western equine encephalitis	Mosquitoes	Wild birds
St. Louis encephalitis virus	St. Louis encephalitis	Mosquitoes	Wild birds
Yellow fever virus	Yellow fever	Mosquitoes	Monkeys
Dengue fever virus	Dengue fever	Mosquitoes	Monkeys

Parasites

Leishmania donovani	Kala-azar (visceral leishmaniasis)	Sandfly	Humans and animals
Leishmania tropica, mexicana, braziliensis	Cutaneous leishmaniasis	Sandfly	Humans and animals
Trypanosoma cruzi	Chagas' disease	Reduviid bug	Humans and animals
Trypanosoma gambiense	West African sleeping sickness	Tsetse fly	Humans
Trypanosoma rhodesiense	East African sleeping sickness	Tsetse fly	Humans
Plasmodium (all)	Malaria	Mosquito	Mosquito

Infections acquired in utero

Treponema pallidum	Congenital syphilis
Cytomegalovirus (CMV)	Cytomegalic inclusion disease
Rubella virus	Congenital rubella syndrome
Toxoplasma gondii	Congenital toxoplasmosis
Listeria	Pneumonia

Infections acquired during passage through the birth canal

Streptococcus agalactiae	Meningitis
Neisseria gonorrhoeae	Ophthalmia neonatorum
Chlamydia trachomatis	Inclusion conjunctivitis Infant pneumonitis
Herpes Simplex II	Neonatal herpes
Candida albicans	Thrush
HBV, CMV, and HIV viruses	

Most common cause of bacterial meningitis

Bacterial meningitis is a medical emergency. Without medical intervention the patient will die within hours. Because of the need for rapid antibiotic treatment it is important to know what the most likely organism is and the best antibiotic for empiric treatment.

Age group	organisms	empiric antibiotic
Neonates (maternal IgG protects against *H. flu*)	1. *Streptococcus agalactiae* (Group B) 2. *E. coli* 3. *Streptococcus pneumoniae* 4. *Neisseria meningitidis* 5. *Listeria monocytogenes*	cefotaxime and ampicillin
6 months-6 years	1. *Haemophilus influenza* type b 2. *Neisseria meningitidis* 3. *Streptococcus pneumoniae*	ceftizoxime cefotaxime
Adults	1. *Streptococcus pneumoniae* 2. *Neisseria meningitidis* 3. *Haemophilus influenza* type b	ceftizoxime cefotaxime
Elderly	1. *Streptococcus pneumoniae* 2. *Neisseria meningitidis* 3. *Haemophilus influenza* type b 4. *Listeria monocytogenes*	ceftizoxime and ampicillin

Most common cause of meningitis in special cases

Epidemics	*Neisseria meningitidis*
Cancer patients	*Listeria monocytogenes*
Renal transplant patients	*Listeria monocytogenes*
Complement deficiency (C6-C8)	*Neisseria meningitidis*
Splenectomy	*Streptococcus pneumoniae* *Haemophilus influenza* Coxsackievirus A and B, Echo virus, Mumps virus *Neisseria meningitidis*
Sickle cell disease	*Streptococcus pneumoniae* *Haemophilus influenza*
Aseptic meningitis	Coxsackievirus A and B, Echo virus, Mumps virus

Interpreting spinal taps for meningitis
The spinal tap is the diagnostic test of choice for meningitis.

pathology	WBC count	neutrophils	protein	glucose
Normal	0	N/A	45 mg/dl	2/3 blood glucose
Bacterial meningitis	>500	> 90% PMNs	> 150 mg/dl	< 40 mg/dl
Aseptic meningitis	50-500	early > 50% late < 20%	< 100 mg/dl	normal
Herpes simplex encephalitis	<1000	< 50%	< 100	normal
TB or cryptococcal meningitis	50-500	early > 50% late < 50%	> 150 mg/dl	< 30 mg/dl
Syphilitic meningitis	50-500	< 10%	< 100 mg/dl	< 40 mg/dl

Notes

Notes

Immunology
The really short course

Immunology has a large cast of characters which allows it to recognize and destroy foreign bodies. These are molecules such as antibodies and receptors and cells such as lymphocytes and macrophages. This quick overview will hopefully keep them straight and be handy when they come up during microbiology, as they often do.

Five antibodies are the principle mechanism of the humoral immune system.

antibody	biological properties	structure	location	notes
IgG	Predominant Ig in serum (15% of serum protein). Very versatile antibody with many actions. • Causes **agglutination reaction** (clumping so that antigens precipitate out of solution). This allows easy phagocytosis. • **Opsonizing activity**. These antibodies enhance phagocytosis when bound to antigens because macrophages and PMNs have receptors for the Fc portion of IgG. • **Antibody dependent, cell-mediated cytotoxicity**. The IgG binds to antigens and concentrates the activity of natural killers lymphocytes (NK cells). • **Activates complement**. • **Neutralizes toxins** by blocking their active sites. • Has **anti-viral activity** by blocking the viral attachment to the cell. • IgG is the predominant antibody in the **second exposure of an antigen** (amnestic response). In the first exposure IgM>IgG but memory cells produce IgG and confer long-lived immunity.	monomer	Vascular and extracellular compartments. Serum half life is 23 days. Crosses the placenta. Found in breast milk.	4 subclasses. Subclass 3 has a short half life (7 days). Subclass 2 cannot cross the placenta. Used in passive immunizations. **Erythroblastosis fetalis.** If mom has IgG antibodies to the fetal Rh group the fetus will get **hemolytic disease of the newborn.** This only occurs if mom is Rh(-) and the child is Rh(+). If the mom does not have the Rh group then she can make anti-Rh antibodies after exposure from a previous Rh(+)child or a blood transfusion. Since IgG passes through the placenta the anti-Rh antibodies will destroy the fetal RBCs.
IgM	**First Ig formed** in an immune response. Increased levels indicate **recent infection.** Antibodies which are formed in response to active immunization. Most efficient **agglutinating** antibody. Most efficient **complement fixing** antibody. Only one molecule is needed to start the cascade rather than the multiple IgG molecules needed to bind complement.	pentamer The five subunits are held together by a **j-chain**.	The pentamer is found primarily in serum A monomer is found on the surface of B-lymphocytes where binding antigen triggers the production of antibodies. Serum half life is 5 days. Fetus begins producing IgM at 5 months.	**Natural isohemagglutinin-antibodies are IgM**. These antibodies are specific for RBC polysaccharides and are responsible for the ABO blood types. Because IgM does not pass through the placenta mother and child can safely have different ABO blood types.

antibody	biological properties	structure	location	notes
IgA	Predominant Ig in secretions. There is more IgA produced than any other Ig. Blocks the attachment of bacteria to mucosal surfaces preventing infection by this route. Agglutination reaction Does **not** activate complement (can activate the alternative complement pathway).	Secreted as a dimer held together by a **j-chain**. An **S component**, created in epithelial cells, is attached to the dimer to allow secretion through the epithelial layer and protect it from proteolysis (digestion) in the GI tract. In serum it is a monomer.	Tears, saliva, sweat, colostrum, urine, gastric fluid and mucus. Serum half life is 5.5 days.	2 subclasses
IgD	No known serum function, is found on the surface of B-lymphocytes and may be important in initiating an immune response or in their maturation.	monomer	Serum half life is 2.8 days.	Easily degraded by proteolytic enzymes.
IgE	An important cause of hypersensitivity reactions. When surface IgE binds an antigen it causes the mast or basophil to release the granules of histamine, heparin and leukotrienes. If this reaction has a detrimental outcome (allergies or anaphylaxis) it is a hypersensitivity reaction. Helps protect against parasites by attracting eosinophils to the parasite.	Monomer with an extra C_H domain. This allows it to bind tightly to basophils and mast cells. This keeps serum concentration low because free IgE is quickly bound to these cells.	Serum half life is 2 days, the shortest of all Ig. Lowest serum concentration of all Ig.	

Hypersensitivity reactions are immunological responses out of proportion to the stimuli that triggered them. These usually result in more damage than good. The hypersensitivity reactions are broken down in to four groups. Each group represents a normal immune response which is over done.

hypersensitivity reaction	mechanism	pathology	examples
Type I **anaphylactic** Mediator: IgE antibody Type I anaphylactic	When IgE on the surface of mast cells and basophils binds an antigen the cells degranulate releasing large amounts of histamine, heparin, leukotrienes and platelet activating factor. To trigger the reaction two IgE molecules must bind one antigen simultaneously (cross-linking).	Mild reactions resemble hay fever: hives, eczema, rhinitis and conjunctivitis. Smooth muscle constriction. Capillary vasodilation and increased permeability leading to edema. If the reaction is systemic it is **anaphylaxis** and can be life threatening. Anaphylaxis is characterized by bronchial constriction and laryngeal edema leading to respiratory failure and vascular collapse secondary to hypotension.	Bee sting allergies. Rare penicillin allergies. In both cases a person must first be sensitized to the antigen to produce the IgE. Treatment is with epinephrine.
Type II **cytotoxic** Mediators: IgG and IgM antibodies Type II cytotoxic	Antibodies which are specific to foreign antigens can sometimes cross react with host tissue or attack foreign antigens attached to host cells. This results in activation of complement, opsonization and other reactions against host cells, autoimmune disease. IgM or IgG antibodies only because it must be able to bind complement.	Complement lysis of cells and chemotaxis of inflammatory cells.	Rheumatic fever. Possibly related to some of the clinical manifestations of Coxsackie infection. ABO transfusion incompatibilities Hemolytic disease of the newborn (erythroblastosis fetalis).

hypersensitivity reaction	mechanism	pathology	examples
Type III **immune complex disease** Mediators: IgG and IgM antibodies	Damage to tissue is the result of the accumulation of antigen-antibody complexes which precipitate out of the blood stream. Most of the time these clumps are cleared by the RES. When these complexes get caught in the capillaries, they imbed in various tissues and activate complement. The complement attacks the antigen and near by "innocent by-standers" (host cells and tissues).	May cause vasculitis. Phagocytes are drawn to the inflammation and release proteolytic enzymes which further damages the tissues. May cause acute or chronic damage depending on how persistent the antigen is.	Acute glomerulonephritis. Systemic manifestations of bacterial endocarditis (arthritis, pericarditis, pleurisy. Serum sickness. Rheumatoid arthritis.
Type IV **delayed type or cell-mediated** Mediator: T-Lymphocytes	Caused by antigens which induce T-cell immunity (primarily mycobacteria and fungi). After initial exposure the body produces CD+4 T cells which recognize the antigen in association with antigen presenting cells. These T-cells release cytokines which draw inflammatory	Mononuclear infiltration and fibrin deposition.	**Contact hypersensitivity** may be induced by poison ivy, soap, topical drugs. Molecules pass through the skin and act as haptens binding to host proteins and sensitizing T-cells to the antigen. On reexposure one gets erythema, itching and necrosis. Skin tests in tuberculosis, leprosy and coccidioidomycosis use delayed hypersensitivity to establish previous or active infection.

Vaccines

Prevention of infectious disease depends on either avoiding exposure or preparing the immune system to eliminate pathogens. Vaccines are the primary means of training the immune system to detect and destroy pathogens.

Vaccination	the administration of any substance intended to provide immunity to future infection. This is inducing the immune system to act as if there has been a prior infection when the patienthas never been infected before.	Giving a vaccination does not mean the patient will become to immune to the disease.
Immunization	The process of becoming immune to a particular infectious disease due to vaccination.	The goal of vaccination is to immunize patients.
Active immunization	inducing the immune system to fight a pathogen by administering a sample of the pathogen either: • a whole killed virus or bacteria • a gentically non-pathogenic but living sample of the pathogen • a part of the pathogen such as a capsule protein • a denatured toxin	
Passive immunization	the administration of exogenously produced antibodies to prevent infection. Often used to protect people from disease after exposure, because active immunization is delayed. Passive immunity wears off when the antibodies are lost or metabolized.	Nature uses passive immunization in the transplacental transfer of IgG to the fetus. Provides passive immunity for the first 4-6 months of life
Toxoid	a bacterial toxin which has been altered so it no longer is toxic but retains its antigenic profile so the body develops neutralizing antibodies to the toxin, called antitoxin.	tetanus toxoid is best example typically produce only short lived immunity requiring frequent boosters.
Immune globulin gamma globulin	A human blood product which contains concentrated antibodies whicha are unsorted giving the recipient humoral protection against a wide range of diseases. Contains 95% IgG and trace amounts of IgM and IgA.	Used in immunocompromised hosts and often in diseases with an autoimmune componant: Guillain Barre, Kawasaki Disease, Immune Thrombocytopenic purpura.
Adjuvents	a chemical given with the antigenic componant of the vaccine to enhance the immune response. Aluminum salts are often used for this purpose.	

The Microbiology Companion, Topf and Faubel ©1997

X-Reference

passive immunizations

Disease	What it is	When to use
Measels	standard human immunoglobulin	Exposure to measels in: immunocompromised hosts, pregnant women or infants under one year. Must be used within 6 days of exposure. After using IVIG one should delay subsequent measles vaccination for at least 5 months.
Tetanus	Special human tetanus immune globulin (TIG)	Used in *dirty* traumatic injuries in patients uncertain of their immunizzation history and patients who have never completed the three primary doses. Also used to treat tetanus (lockjaw).
Rabies	Special human rabies immunoglobulin (RIG) is preferred form	In all unimmunized people bit by a high risk animal (bats, racoons) or by an unprovoked attack from an animal in which rabies cannot be ruled out. The rabies killed vaccine is given with the human immunoglobulin.
Hepatitis B	Special human hepatitis B immune globulin (HBIG)	Recommended for infants born to mothers with active hepatitis B. Also used in post exposure prophylaxis in unimmunized individuals. Always given with HBV, in different sites.
Varicella zoster	Varicella zoster immunoglobulin (VZIG)	Used to prevent disseminated zoster in immunocompromised patients who have been exposed.
CMV	CMV-IGIV	Has been used in immunocompromised patients along with gancyclovir to treat CMV pneumonia. Has been tested in CMV – transplant patients.
RSV	RSV-IGIV	Recommended for children under two years old with bronchopulmonary dysplasia (BPD, lung scarring associated with prematurity) or a history of prematurity (earlier than 32 wks). Given monthly during RSV season.
B-19 parvovirus	human immunoglobulin	Has been used with some success in immunodeficient patients with chronic B-19 infection
Botulism	Equine anti-toxin for types A, B and E Human derived botulism antitoxin is undergoing trials for infants.	Used in food-borne and wound botulism, should be given to the patient as soon as possible. Not recommended for infant botulism.

Routine Childhood (active) Vaccines

Vaccine	organism	content of vaccine	notes
DTaP or DTwP	*Bordetella pertussis*	aP is an acellular vaccine which is the recommended form. contains purified proteins derived from the organism wP is an inactivated bacteria	the acellular vaccine has a lower risk of systemic side effects, including seizures. Do not give to children older than 7 due to a high rate of complications. Use Td instead. fever is a common side effect
	Clostridium tetani	Tetanus toxoid (inactivated toxin).	encephalopathy has been reported in 1 in 300,000 doses
	Corynebacterium diphtheriae	dipheria toxoid exotoxin denatured with formaldehyde	Kids who have had a reaction to the pertussis componant should complete immunization with TD. After age 7 give Td, low dose diphtheria.
HiB	*Haemophilus influenza*	Polysaccharide capsular antigen of type b conjugated (attached) to a protein carrier.	Tetramune is HiB combined with DTwP, Hib with DTaP will soon be licensed for all four doses.
HBV	Hepatitis B	Synthetic recombinant vaccine made from yeast cells. Contains HBsAg.	Recommended for adults at risk: health care workers, IVDA, gay males, multiple sex partners. HBsAb appears in serum.
TOPV	Poliovirus	Live attenuated virus (Sabin)	Considered alternative to recommended IPV for first two doses. Do not give to kids with immunodefficiency or with close contacts to people with impaired immunity.
IPV	Poliovirus	killed virus. (Salk)	Polio vaccine of choice due to the lowered prevalence of wild type virrus

Vaccine	organism	content of vaccine	notes
MMR	Rubella Measles Mumps	Live attenuated virus	The viruses are incubated in chicken eggs and are contraindicated in patients with anaphylactic reactions to eggs. Pregnancy is a contraindication to MMR due to the birth defects associated with rubella. Women who have received the vaccine by accident should be councled to the risk to the fetus. No documented cases of vaccine associated birth defects are available
Var	Varicella	live attenuated virus	One dose for kids under age 12 and two doses for teenagers and adults.

vaccine	birth	1 mo	2 mo	4 mo	6 mo	12 mo	15 mo	4-6 yrs	11-12 yrs	13-16 yrs
HBV	███	███			███				teens at high risk need the hepatitis series *	
DTaP			███	███	███		███	███		
DT									███	
Hib			███	███	███					
IPV			███	███		███		███		
OPV			†			███				
MMR						███		**OR** ███	███	
Var						███ **OR** ███			**§**	**‡**

* Only adolescents who were unimmunized as a child need to get the HBV series. As of now their are no recommendations for a booster series though the need for one is suspected

§ Varicella may be given to children who have not acquired natural immunity by age 11.

‡ If the varicella vaccine is given after age 13 two doses given one month apart are required for reliable immunity.

† Currently the following regiments are deemed acceptable by the ACIP, AAP and the AAFP:
1. IPV at 2 and 4 months; OPV at 12-18 months and 4-6 years
2. IPV at 2, 4 ,12-18 months and 4-6 years
3. OPV at 2, 4 ,12-18 months and 4-6 years

The advisory council on immunization Practices (ACIP) recommends schedule 1.

Other Active Vaccines

organism name of vaccine	content of vaccine	recipients of vaccine
Adenovirus	Live attenuated virus (serotypes 4 and 7).	Military recruits. Given orally.
Bacillus anthracis Anthrax vaccine	Cell free vaccine Only short term protection and boosters must be given annually.	Available for persons at risk: people who work with animals or animal hides. Distributred by Michigan Dept of Health
Francisella tularensis	Live attenuated virus.	Only for persons at risk (i.e. workers in contact with animals).
Influenza	Inactivated virus which must be updated yearly to match the prevalent strains.	Persons with a chronic heart or lung condition. Persons over the age of 65. Intranasal version in development.
Hepatitis A	Purified viral anitgens.	Persons traveling to or living in endemic areas.
Japanese encephalitis mosquito borne arbovirus	inactivated virus	Vaccination recommended for people living in endemic east Asia and travelers staying longer than a month.
Mycobacteria tuberculosis BCG (bacillus Calmette-Guerin)	Live attenuated bacteria. BCG is used in the treatment of bladder cancer to decrease tumor recurrence.	Rarely used in the U.S. Can be considered for use in populations with high conversion rate and poor access or compliance with medical care.
Neisseria meningitidis	Polysaccharide capsule of A, C, Y and W-135 antigenic types. Note absence of B serotype.	Military recruits, asplenic, compliment deficiency. Outbreak control in individuals in close contact with infected persons.
Rabies virus	Inactivated attenuated virus cultured from human cells.	People at risk from occupational exposure such a veterinarians. Post-exposure prophylaxis along with rabies immune globulin (separate sites).
Rickettsia prowazekii (typhus fever)	Inactivated bacteria.	Persons traveling to endemic areas or exposed to someone who has traveled to an endemic area.

Salmonella typhi (typhoid)	Three varieties:
	  • inactivated bacteria
	  • capsular polysaccharide
	  • live vaccine (given orally)
		Persons traveling to endemic areas or exposed to a patient with a confirmed case. Disaster relief.

organism name of vaccine	content of vaccine	recipients of vaccine
Smallpox virus	Live attenuated virus.	Worldwide use of the vaccine led to the eradication of the disease in the late 1970s. Now only military personnel are vaccinated.
Streptococcus pneumoniae pneumococcal vaccine	Polysaccharide capsule of 23 antigenic types.	Persons without a functional spleen or with a chronic heart or lung condition. Persons over the age of 65.
Yellow fever virus	Live attenuated virus.	Persons traveling to or living in endemic areas.
Yersinia pestis (plague)	Inactivated bacteria.	Agricultural workers.
Vibrio cholera	Inactivated bacteria.	Ineffective vaccine. Protects only half of the recipients and then for only 3–6 months. Only used in contacts of confirmed cases and in controlling outbreaks. Water precautions are more important for prevention than the vaccine.

X-Reference

Notes

X-Reference

acid-fast. The ability of an organism to retain a stain when subjected to a wash of 95% alcohol with 3% HCl. This alcohol-acid solution is able to decolorize all non-acid-fast organisms. *Mycobacterium* and *Nocardia* are examples of acid-fast organisms. Acid-fast organisms are stained using the Ziehl-Neelsen technique. (see also: Ziehl-Neelsen)

alpha hemolysis. Represents incomplete hemolysis of red blood cells on a blood agar plate resulting in a greenish brown haze surrounding the colony. (see also: sheep blood agar)

anicteric. Without jaundice.

arthropod. Any member of the phylum Arthropoda which includes centipedes, crustaceans, insects, millipedes, mites, scorpions, spiders and other species. Arthropods represent about 75% of known animal species. Arthropods, particularly insects, can serve as vectors for human disease.

atypical pneumonia. Any lung infection whose causative agent cannot be isolated on ordinary culture media or when the clinical course does not resemble pneumococcal pneumonia. Typically caused by a virus or *Mycoplasma*.

trait	**typical** pneumonia	**atypical** pneumonia
onset	acute	gradual
chills	+	-
pleuritic pain	+	-
cough	productive	dry
fever	greater than 38.9°C (102°F)	less than 38.9°C (102°F)
distribution	one lobe or segment	multiple lobes
typical pathogen	*Strep pneumonia*	*Mycoplasma*

bacteriophage (syn. phage). A virus which infects bacteria. (see also: lysogeny).

beta hemolysis. Represents complete hemolysis of red blood cells on a sheep blood agar plate resulting in a clear zone or halo surrounding the colony. (see also sheep blood agar)

blood agar plate. See sheep blood agar.

bronchiolitis. Inflammation of the bronchioles which is often associated with bronchopneumonia.

bronchopneumonia. An acute inflammation of the smaller bronchial tubes which results in the deposition of a mucopurulent exudate in the bronchi causing an irregular pattern of consolidation.

X–Reference

budding. A form of asexual reproduction in which the organism divides into two unequal parts. The larger part is considered the parent and the smaller part the progeny. Yeasts divide in this manner. Compare with fission.

chocolate agar plate. A nonselective culture medium which is enriched with heated sheep blood. Heating the blood releases red cell components such as hemoglobin and nicotinamide adenine dinucleotide (NAD) which are required for the growth of *Haemophilus*.

coliform. Bacteria that live in the GI tract which normally do not cause disease. They can cause disease when they get into other parts of the body or acquire plasmids containing virulence factors.

core polysaccharide. See lipopolysaccharide.

cross reaction. Occurs when antibodies against a particular antigen bind with antigens other than the ones that induces them.

cyst. 1. An abnormal membrane bound sac containing fluid or gas. 2. Urinary bladder. 3. The durable form of a parasite which is able to survive outside the host.

cystitis. Inflammation of the urinary bladder.

disseminated intravascular coagulation. A pathologic condition in which both clotting factors and fibrinolytic enzymes are activated. Coagulation of blood and lysis of clots occur throughout the body. The final result is tissue necrosis and hemorrhage. This phenomenon may occur during gram negative septicemia due to endotoxin.

ELISA (enzyme linked immunosorbent assay) a diagnostic technique in which:

An enzyme is attached to an antibody. The antibody can be against any particular protein (antigen).

The antibody-enzyme complex is then added to the sample (often serum) which is being tested.

This mixture is then rinsed so that only antibody-enzyme complex linked to the antigen remains.

Then, substrate for the enzyme is added. The substrate and the enzyme react to produce color. By determining the presence and/or intensity of the color, the presence/ amount of product (antigen) can be determined.

Early pregnancy tests and the screening test for HIV use ELISA. For pregnancy tests the protein antigen is HCG (human chorionic gonadotropin); for the HIV screening test the protein antigen is antibody to HIV.

fission. A form of asexual reproduction where the mother cell divides into two daughter cell of equal size. Bacteria divide in this manner. Compare with budding.

fomite. An inanimate vector of disease. For example, impetigo (caused by *Streptococcus pyogenes*) may be transmitted when a person uses a towel used by an infected person. The towel harbors the organism and is a fomite.

gangrene. Tissue necrosis which is usually secondary to poor circulation or infection.

genome. The complete genetic record of a cell or organism. It is coded in the DNA of all eukaryotic and prokaryotic cells. The genome of a virus may be coded as DNA or RNA.

Gram staining. A staining technique for bacteria developed by Christian Gram in 1884. Organisms which are gram positive stain blue and those which are gram negative stain red/pink. The technique is as follows:

1. Stain with crystal **violet**. All bacteria will stain blue with this stain.

2. Stain with potassium iodide-**iodine** solution.

3. Add decolorizer: **alcohol**. Until this step, all bacteria react the same way. After decoloration, gram positive organisms remain blue but the stain is removed from gram negative organisms.

4. Stain with counterstain: **safranin**. Stains gram negative organism pink/red.

hemolysis. Refers to the lysis/destruction of red blood cells. Some bacteria when grown on sheep blood agar release hemolysins which cause hemolysis of the sheep RBCs contained in the agar. Three types of hemolysis are described; alpha: incomplete hemolysis (appears green), beta: complete hemolysis (appears clear) and gamma: no hemolysis.

hepatomegaly. Denotes an abnormally large liver.

host. Refers to any cell or organism which harbors another organism A parasite needs a host to live.

iatrogenic infection. An infection which is caused by the treatment of a patient. Compare to nosocomial.

---

immunocompetent. Refers to an individual whose immune system is functioning normally.

immunocompromised. Refers to an individual whose immune system is not working 100%. This includes burn victims who have lost skin, children with congenital immune defects, and persons with leukemia or AIDS.

Note how similar these two words are. Don't make a stupid mistake by mixing these up on a test. Stop and read carefully any time these words show up.

---

incubation period. The period of time between infection and the first symptoms.

infection. Invasion of the body by pathogenic organisms such as bacteria, viruses, fungi and protozoa.

L-forms. Any bacteria which has lost the ability to produce a cell wall but can still reproduce. They do not cause disease in humans. L-forms are selected for when an infection is treated with cell wall antibiotics which have no effect against L-forms; after the treatment has stopped, the L-forms have the potential to revert back to the virulent form (able to produce a cell wall) and cause relapse of disease.

Lipid A. See lipopolysaccharide.

lipopolysaccharide (LPS). A component of the gram negative cell wall which is unique to gram negative bacteria. It is a glycolipid which makes up the lipids in the outer leaf of the outer membrane. It is toxic to humans and animals and many of the symptoms of gram negative infection can be replicated with purified LPS. LPS is made up of three components:

Lipid A is a phospholipid which anchors the LPS in the outer membrane. It varies among species but is constant within a species. This is the toxic component.

Core polysaccharides are a short train of sugars attached to the lipid and are constant within a genus.

O-specific side chain is a hypervariable region consisting of a backbone of 3-5 sugars which is repeated about 50 times. It varies within species and is used to serotype species. O157:H7 designates *E. coli* O side chain serotype 157 and flagella serotype 7.

lobar pneumonia. An acute febrile illness marked by inflammation of one or more lobes of the lung. The inflammation is characterized by consolidation, rusty sputum and dyspnea (shortness of breath). It is most often seen in the lower lobes. 90% of lobar pneumonia is caused by *Streptococcus pneumoniae*.

Löwenstein-Jensen medium. An egg-based culture medium used to culture all species of *Mycobacteria*.

lymphadenopathy. Any disease effecting the lymph nodes. Usually associated with nodes which are swollen and tender.

lysogeny. Occurs when a bacteriophage invades a bacteria. The phage incorporates its DNA into the bacteria's genome and remains latent. At a later time the phage DNA will become activated, produce progeny and lyse the host cell. Often the phage DNA codes for a toxin or virulence factor for the bacteria; in this case, only bacteria which are infected by the phage can cause disease.

MacConkey agar. A selective and differential culture medium for growing gram negative rods. The growth of gram positive bacteria is inhibited by bile salts and crystal violet contained in the agar. The agar also differentiates gram negative rods into lactose fermenters and non-lactose fermenters. **Colonies which ferment lactose appear red or pink**.

meningitis. An inflammation of the meninges (membranes) of the brain and spinal cord.

aseptic meningitis. Meningitis which is associated with an increase of lymphocytes and monocytes in the CSF and is usually caused by a virus. The term aseptic means: "without living pathogenic organisms," this misnomer is used to describe this type of meningitis because the cause (viral) is not visible and cannot be cultured like bacteria or fungi.

purulent meningitis. Meningitis which is associated with an increase of PMNs in the CSF and is usually caused by bacteria or fungi.

nosocomial infection. An infection which is secondary to a hospital visit. Compare to iatrogenic.

nucleocapsid. A protein coat which surrounds the viral genome and other viral proteins (reverse transcriptase, RNA dependent RNA polymerase). It can be one of two morphologies:

icosahedral   helical

O-specific side chain. See lipopolysaccharide.

orphan. A virus which is not associated with any disease. Echovirus and Reovirus are both acronyms with the O standing for orphan. This is a misnomer because both of these viruses cause diseases.

osteomyelitis. Infection of bone marrow and surrounding bone. *Staphylococcus aureus* is the organism most frequently responsible.

oxidase positive. Indicates that an organism can oxidize dimethylparaphenylene and tetramethylparaphenylene (editor's note: the two biggest words in the book). This turns the colony black. Any organism with cytochrome oxidase in its respiratory chain will be oxidase positive.

pandemic. A widespread epidemic.

phage. See bacteriophage.

phase variation. A change in the antigenic structure of a pathogen which allows it to avoid the body's immune response. *Salmonella* uses phase variation to change between flagella H1 and H2.

Phase variation of *Salmonella* is possible because the promoter for H2 is able to invert. In its natural state, the promoter for H2 activates the transcription of H2 and the repressor for the H1 gene; hence expression of H2 concomitantly prevents the expression of H1. When the H2 promoter inverts, it cannot promote H2 or the H1 repressor.

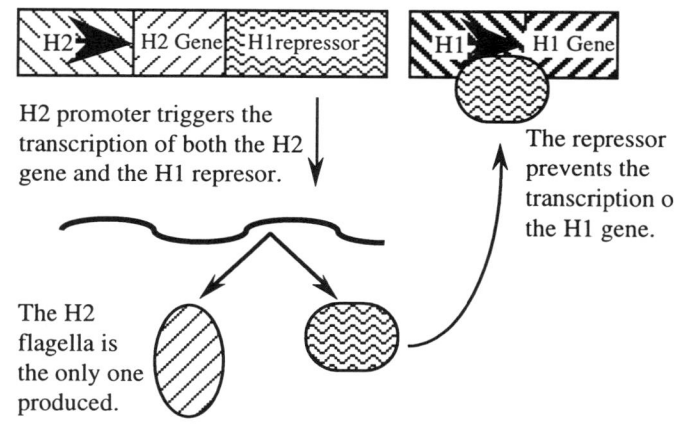

H2 promoter triggers the transcription of both the H2 gene and the H1 represor.

The repressor prevents the transcription of the H1 gene.

The H2 flagella is the only one produced.

The H2 promoter is inverted once every 1000 generations and can no longer promote the H2 gene or the repressor gene.

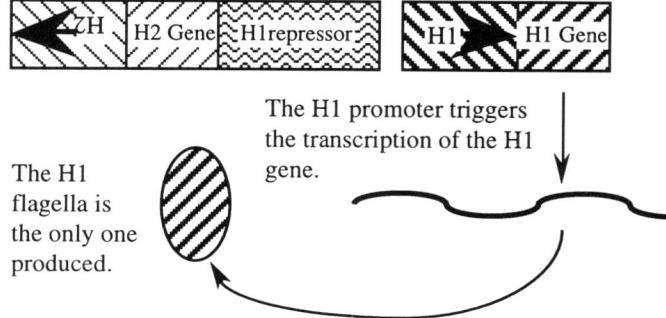

The H1 promoter triggers the transcription of the H1 gene.

The H1 flagella is the only one produced.

Glossary

plasmid. An extrachromosomal piece of DNA found in bacteria. It often contains genes that confer antibiotic resistance. Plasmids are readily spread among a population of bacteria while chromosomal DNA is not.

pneumonia. Inflammation of the parenchyma of the lungs. Lung parenchyma includes tissue beyond the bronchi. Affected areas become consolidated when the air spaces fill with blood cells and fibrin.

pruritic. Itchy.

reservoir. A host which harbors a pathogen without getting the disease. This allows the pathogen to survive and infect others. Reservoirs are important for the perpetuation of arthropod borne diseases because winter can kill all of the infected insects. The next generation of insects can then pick up the disease from a mammalian reservoir which survived through the winter.

reverse transcriptase. A viral encoded RNA dependent DNA polymerase. It is the defining characteristic of retroviruses. Reverse transcriptase transcribes viral RNA into DNA which is inserted in the host genome. It is called "reverse" because it transcribes in the opposite direction of the normal flow of information which is from DNA to RNA. Note that hepatitis B also has a reverse transcriptase.

saprophyte. An organism which eats dead plant and animal matter.

self assemble. Refers to the ability of viruses assemble their nucleocapsids without any energy input. The organization of the capsid proteins is spontaneous.

serum sickness. May occur when foreign antigens (i.e. horse immunoglobulins) are injected into the body. The body produces antibodies against the antigens and the antibody-antigen complexes deposit all over the body resulting in arthritis, nephritis, rash and fever. Type III hypersensitivity reaction (this is also known as immune complex hypersensitivity).

splenomegaly. Denotes an abnormally large spleen.

sheep blood agar. A general purpose culture medium for growing bacteria. It is particularly useful for distinguishing Streptococcal species because it demonstrates hemolytic properties.

Thayer-Martin medium. A selective culture medium that consists of chocolate agar with colistin (inhibits gram negative bacilli), vancomycin (inhibits gram positive cocci), and nystatin (inhibits fungi) which suppress the growth of normal flora. It is use to culture *Neisseria meningitidis* and *gonorrhoeae*.

toxoid. A non-toxic derivative of a toxin which can be used to immunize a person against the toxin.

transovarian transmission. Occurs when insect vectors of disease which are able to pass the pathogen on to its progeny. This makes the insect both a vector and a reservoir.

type specific immunity. Immunity which develops against an antigen which has numerous serotypes. Immunity which develops against one serotype is not protective when a person is infected with another serotype. So in the case of *Streptococcus pyogenes* antibodies against the M protein are protective, however there are over 60 different M proteins. Therefore, it is possible to be reinfected with *Streptococcus pyogenes* several times; hence, the immunity is type specific.

vector. A carrier of disease which transmits a pathogen from an infected individual to another host. Insects often serve as vectors.

Ziehl-Neelsen staining. A staining technique which is used to stain acid-fast bacteria. Acid-fast bacteria do not stain with routine gram staining. The technique is as follows:

1. Stain with the red stain carbolfuchsin and heat for several minutes.

2. Decolorize with acid-alcohol solution. Bacteria which are acid-fast will not decolorize.

3. Counterstain with methylene blue. Organisms which are not acid-fast will take up the stain and appear blue; acid fast organisms retain the carbolfuchsin stain and appear red.

zoonoses. Diseases of humans for which animals are the reservoirs.

Index

Index

Acquired immunodeficiency syndrome **147**
Actinomyces israelii **79**
Actinomycosis **79**
Acute glomerulonephritis **7**
acute respiratory disease **121**
Acute rheumatic fever **7**
acyclovir **109**
Adenoviridae **120, 121**
Adenovirus **121**
ADP/ATP translocators **67**
African eye worm **222**
African trypanosomiasis **185**
agglutination **241**
Air conditioners **52**
alpha genes **106**
Alpha hemolysis **227**
Alpha toxin **4, 24**
Altered penicillin binding proteins **4**
Amastigotes **182**
Amebiasis **179**
Amebic dysentery **179**
anaphylaxis **242, 243**
Ancylostoma braziliense **218**
Ancylostoma caninum **218**
Ancylostoma duodenale **212**
anergy **82**
Anopheles mosquito **187**
antelopes **185**
Anthrax **17**
Anti Streptolysin O (ASO) **6**
Antibody dependent, cell-mediated
 cytotoxicity **241**
Antigenic
 drift **126**
Antigenic shift **126**

Aplastic crisis **123**
arboviruses **134**
Ascaris lumbricoides **211**
Aseptic meningitis **237**
aspergillosis **173**
Aspergillus fumigatus **173**
athletes foot **175**
Atypical Mycobacteria **87**
Atypical pneumonia **68, 78**
AZT **149**
B19 **123**
Bacillus **17**
 anthracis **17**
 cereus **17**
bacitracin **6, 9, 10, 11**
Bacterial meningitis **236, 237**
Bacteroides **48**
 fragilis **48**
basophils **242**
bat guano **169**
BCG vaccine **82**
bear meat **217**
beef tapeworm **196**
beta genes **106**
Beta hemolysis **227**
beta-lactamase **4**
BK virus **118**
Black Death **56**
Blackwater fever **188**
bladder cancer **204**
Blastomyces dermatitidis **167**
Blastomycosis **167**
body louse **59**
Bordet-Gengou medium **52**
Bordetella pertussis **52**
Borrelia **59**
 burgdorferi **59**

 recurrentis **59**
bridging necrosis **97**
Brill-Zinsser disease **74**
Brucella
 abortus **54**
 melitensis **54**
 suis **54**
Brucellosis **54**
Brugia malayi **221**
buboes **56**
Bubonic Plague **56**
Burkitt's lymphoma **114**
C carbohydrate **6**
C-100-3 antigen **138**
cachexia **80**
Calabar swellings **222**
Campylobacter jejuni **46**
Candida albicans **171**
capsule **226**
cat hookworm **218**
catalase + **4**
Catalase positive **228**
cercariae **203**
Cervical carcinoma **109, 120**
cervicitis. **70**
Chagas' disease **184**
Chancroid
 Haemophilus ducreyi **51**
chickenpox **111**
Chlamydia **67**
 pneumoniae **68**
 psittaci **68**
Chlamydiae **70**
 trachomatis **69**
chlamydospores **171**
choleragen **44**
cholesterol **78**

Chorioamnionitis 9
chorioretinitis 190
chromatoidal bars 179
Chronic fatigue syndrome 114
Cladosporium werneckii 176
clindamycin 24
Clonorchiasis 205
Clonorchis sinensis 205
Clostridia 19
Clostridium
    botulinum 20
    difficile 24
    perfringens 24
    tetani 22
Coccidioides immitis 167
Coccidioidomycosis 167
Coliform 41
Colorado Tick Fever (CTF) 160
colostrum 242
common cold 158
Common warts 120
complement 241
Condylomata acuminata 120
Congenital rubella syndrome 133
conidia 228
Conjunctivitis 51
Contact hypersensitivity 244
copepod 198
coracidia 198
Corynebacterium diphtheriae 26
Coxiella burnetii 76
coxsackievirus 157
crab meat 206
crustacean 198
Cryptococcosis 172
Cryptococcus neoformans 172

Cryptosporidiosis 186
Cryptosporidium 186
culture 229
"currant jelly" sputum 41
cystic fibrosis 43
cysticercosis 197
cysts 178, 179
Cytomegalic inclusion disease 112
cytomegalovirus 112
cytotoxic 243
D-glutamate 17
dairy products
    Brucella 54
    Campylobacter jejuni 46
    Listeria 27
Dane particles 95
darkfield microscopy 63
deer fly 222
deer tick 59
Dengue fever 136
    dengue hemorrhagic fever (DHF) 136
    dengue shock syndrome (DSS) 136
Dental carries 10
Dermacentor andersoni tick 73
Dermacentor variabilis tick 73
dermal nodule 221
Dermatophytid reaction 175
DFA-TP 63
Diarrhea 233
Dieterle silver impregnation 52
Dimorphic 165, 228
Diphtheria 26
Diphtheria toxin 26
Diphyllobothriasis 198
Diphyllobothrium latum 198
dog hookworm 218

dog tapeworm 199
DPT 23
Dysentery 38
East African sleeping sickness 185
Eastern equine encephalitis 134
Echinococcus granulosus 199
Echovirus 157
Edema factor 17
EF-2 43
elementary body 67
elephantiasis 219
elongation factor 2 43
Encephalitis 108
endocarditis, culture negative 76
endoflagella 58
Entamoeba histolytica 179
Enterobacter cloacae 41
Enterobius vermicularis 209
enterovirus 72 158
env 143, 144
eosinophilia 217
Epidermodysplasia verruciformis 120
Epiglottitis 50
epithelioid cells 81
Epstein Barr Virus 114
erythema chronicum migrans 59
Erythema infectiosum 123
Erythroblastosis fetalis 241
Escherichia coli 36
    Enterohemorrhagic 37
    Enteroinvasive 37
    Enteropathogenic 37
    Enterotoxigenic 37
Exfoliative 4
exotoxin 232
Exotoxin A 43

Index

factor V
NAD **50**
factor X
heme **50**
Farmer's lung **173**
febrile seizures **115**
Fifth disease **123**
filariform larvae **213, 218**
fish tapeworm **198**
flavivirus
Dengue fever virus **136**
Saint Louis encephalitis virus **135**
Yellow fever virus **135**
foscarnet **116**
four F's of transmission **38**
Francisella tularensis **55**
freshwater snails **203**
FTA-ABS **64**
Fungus balls **173**
Fusion protein **131**
Fusobacterium **48**
gag **143, 144**
gamma genes **106**
Gas gangrene **24**
General paresis **62**
Genital herpes **109**
Genital tract infection **70**
genital warts **120**
genotypes **120**
Ghon complex **81**
Giardia lamblia **180**
Giardiasis **180**
Giemsa stain **187, 191**
Gilchrist's disease **167**
Gingivostomatitis **108**
gonococcus **13**
Gram-Weigert stains **191**

granuloma **81, 226**
guinea fire worm **220**
gummas **62**
Haemophilus **50**
aegyptius **51**
ducreyi. **51**
influenza **50**
Halophilic **44**
Hand-foot-and-mouth disease **157**
HBcAb **99**
HBcAg **99**
HBeAb **99**
HBeAg **95, 99**
HBsAb **99**
HBsAg **95, 99**
Heat labile toxin (LT) **37**
Heat stable toxin (ST) **37**
Helicobacter pylori **47**
hemagglutinin **121, 128**
hemolysis (beta or alpha hemolytic) **227**
hemolytic disease of the newborn **241**
Hemorrhagic cystitis **121**
Hepadnavirus **95**
heparin **242**
Hepatitis **94-99, 138, 158**
A Virus **158**
acute **97, 138, 158**
C virus **138**
chronic **97, 138**
delta virus **94, 101**
E virus **94, 138**
enterovirus 72 **158**
fulminant **97**
non A-non B **138**
hepatocellular carcinoma (HCC) **97**
Herpangina **157**
Herpes labialis **108**
Herpesviridae **106-116**

herpes simplex virus I **108**
herpes simplex virus II **109**
Herpetic whitlow **108**
histamine **242**
Histoplasma capsulatum **169**
histoplasmoma **169**
Histoplasmosis **169**
Hodgkin's disease **115**
Human Herpesvirus **115**
Human herpesvirus–6 **115**
Human herpesvirus–7 **115**
Human herpesvirus–8 **116**
Human Immunodeficiency Virus
HIV **146**
Human papilloma virus **120**
Human T-cell leukemia virus **153**
HTLV-II **153**
Hydatid cyst disease **199**
Hydrophobia **139**
hypersensitivity reaction **242, 243**
hyphae **165**
hypnozoites **188**
I do have a life outside of medical school
contest **222**
IgA **242**
IgA protease **228**
IgD **242**
IgE **242**
IgG **241**
IgM **241**
immune complex disease **244**
impetigo **7**
Inclusion **69**
India ink test **172**
Infant botulism **20**
Infant pneumonitis **70**
infectious genome **132, 133**

Influenza **126**
INH **81**
initial body **67**
interstitial pneumonia **116**
intracellular **226**
iota toxin **24**
Ixodes dammini **59**
j-chain **241**, **242**
Jarisch-Herxheimer reaction **62**
JC virus **118**
jock itch **175**
Kaposi's sarcoma **116**
Kaposi's Sarcoma **148**
Keratoconjunctivitis **108**
    epidemic **121**
Klebsiella pneumonia **41**
Koplik spots **129**
Lancefield Group **6**
Langhans cells **81**
Legionella pneumophila **52**
Legionnaires' disease **52**
Leishmania **182**
    braziliensis **182**
    donovani **182**
    mexicana **182**
    tropica **182**
leonine facies **86**
Lepromatous leprosy **86**
lepromin skin test **86**
Leptospira interrogans **60**
Leptospirosis **60**
Lethal factor **17**
leukemia **115**
Leukemias and lymphomas **153**
leukotrienes **242**
Listeria monocytogenes **27**

liver fluke **205**
Loa loa **222**
Lobar (typical) pneumonia **10**
lockjaw **22**
Loeffler medium **26**
Loeffler's syndrome **218**
Löwenstein-Jensen medium **82**
lung fluke **206**
Lyme disease **59**
Lymphogranuloma venereum **70**
lysogenic phage **26**
Lytic cycle **118**
M protein **6**
Malassezia furfur **176**
mast cell **242**
Measles virus **129**
megaloblastic anemia **198**
Membrane Protein II **13**
meningitis **236**
Meningoencephalitis **111**
merozoites **186**
Metachromatic staining **26**
methylene blue **38**
microcephaly **112**
microconidia **169**
miracidia **206**
mononucleosis
    heterophile-negative **112**
    Heterophile-positive **114**
motile **226**
Mouse polyoma virus **118**
mu toxin **24**
multinucleated giant cells **81**, **129**
multiple sclerosis **153**
Mumps virus **131**
Murine toxin **56**

myc **114**
Mycobacteria **80-88**
    avium-intracellulare complex **88**
    fortuitum complex **88**
    kansasii **88**
    leprae **86**
    marinum **88**
    scrofulaceum **88**
    tuberculosis **81**
Mycoplasma pneumoniae **78**
Myocarditis **157**
Nasopharyngeal carcinoma **114**
Necator americanus **212**
nef **145**
Negri bodies **139**
Neisseria **12**
    gonorrhoeae **13**
    meningitidis **12**
Neonatal herpes **109**
Neuraminidase **126**
Neutralizes toxins **241**
new world hookworm **212**
Nocardia asteroides **79**
Nocardiosis **79**
non-Hodgkin's lymphoma **115**
Non-O serotype **44**
Nonchromogens **87**
nongonococcal urethritis **70**
Nonpermissive cycle **118**
normal flora **227**
novobiocin **3**
old world hookworm **212**
Onchocerca volvulus **221**
operculum **198**
Ophthalmia neonatorum **13**
Opsonizing activity **241**
optochin **10**

Index

orbivirus **160**
Orchitis **131**
Oriental liver fluke infection **205**
Osteomyelitis **4**
otitis externa **43**
Paracoccidioides brasiliensis **169**
Paragonimiasis **206**
Paragonimus westermani **206**
Paramyxovirus **128**
parotids **131**
Parvoviridae **123**
PCP **191**
pelvic inflammatory disease (PID) **13**
penicillin binding protein **4**
penton **121**
Pertussis toxin **52**
Pharyngitis **6**
Photochromogens **87**
piecemeal necrosis **97**
Pili **6**
pink eye **51**
pinworm **209**
placenta. **241**
Plague **56**
Plasmodium **187**
    falciparum **188**
    malariae **188**
    ovale **188**
    vivax **188**
plerocercoid **198**
Pleurodynia **157**
Pneumocystis carinii pneumonia **191**
pneumonic bubonic plague **56**
pol **144**
Poliomyelitis **155**
Poliovirus **155**
Pontiac fever **52**

pork tapeworm **197**
portal hypertension **203**
Post-herpetic neuralgia **111**
Postpartum endometritis **9**
potassium tellurite **26**
Poxviridae **103-104**
    smallpox **104**
    vaccinia **104**
PPD **82**
Prevotella
    melaninogenica **48**
procercoid **198**
Progressive multifocal
    leukoencephalopathy **118**
Protective antigen **17**
Protein A **4**
Proteus **41**
pseudohyphae **171**
pseudomembrane **26**
Pseudomembranous colitis **24**
Pseudomonas aeruginosa **43**
pseudopods **179**
Psittacosis **68**
pyocyanin **43**
Pyoderma **7**
pyoverdin **43**
Q fever **76**
Rabies virus **139**
Rapid growth **87**
rapid plasma reagin **64**
recurrent fevers **228**
Reduviid bug **184**
Reheated fried rice **17**
Relapsing fever **59**
renal stone **41**
reovirus **160**
reticulate body **67**

rev **145**
reverse transcriptase **95, 143, 144**
Reye's syndrome **111, 126**
rhabditiform larvae **212, 213**
rhabdovirus **139**
    Rabies virus **139**
Rhinocerebral disease **173**
rhinovirus **158**
Riboflavin deficiency **171**
rice water stools **44**
Rickettsiae **72, 75-76**
    akari **73**
    prowazekii **74**
    rickettsii **73**
    typhi **75**
Rickettsialpox **73**
rifampin **12**
ringworm **175**
river blindness **221**
RNA dependent DNA polymerase **95, 143, 144**
Rocky Mountain spotted fever **73**
Rose spots **39**
roseola **115**
rotavirus **160**
RPR **64**
Rubella virus **133**
S component **242**
Sabin Vaccine (TOPV) **155**
safety pin **56**
Saint Louis encephalitis **135**
Salk Vaccine **155**
Salmonella
    cholera suis **39**
    paratyphi **39**
    typhi **39**
Scalded skin syndrome **4**

Scarlet Fever **6**
Schistosoma **203, 204**
Schistosoma japonicum **204**
Schistosomiasis **203**
schizogony **186**
schizont **186**
Scotch Tape™ **209**
Scotchromogens **87**
seafood **44**
Serratia marcescens **41**
Shiga toxin **38**
Shiga-like toxin **37**
Shigella
    boydii **38**
    dysenteriae **38**
    flexneri **38**
    sonnei **38**
Simian Vacuolating Virus **118**
Sleeping sickness **185**
slow lactose fermenter
    Serratia marcescens **41**
    sonnei **38**
    Vibrio **44**
Somalia **104**
spherules **167**
spinal tap **237**
Spirochetes **58-59**
sporogony **186**
Sporothrix schenckii **176**
Sporotrichosis **176**
Staphylococcus **3**
    aureus **4**
    epidermidis **3**
    saprophyticus **3**
Streptococcus **6-10**
    agalactiae **9**
    bovis **11**

pneumoniae **10**
pyogenes **6, 7**
viridans **10**
Streptokinase **6**
Streptolysin O **6**
Streptolysin S **6**
Strongyloides stercoralis **213**
Subacute sclerosing panencephalitis (SSPE) **129**
sulfur granules **79**
SV40 **118**
Swimmer's ear **43**
Swimmers itch **203**
Syphilis **61**
T-antigens **117**
Taenia saginata **196**
Taenia solium **197**
Tag **117**
tax **145**
Td **23**
Tertiary syphilis **62**
Tetanus **22**
tetanus immune globulin **23**
Tetanus Immunization **23**
Thayer-Martin medium **12, 13**
thickened skin **221**
thrush, **171**
TIG **23**
Tinea **175**
Tinea nigra **176**
Tinea versicolor **176**
Toxic shock syndrome **4**
toxin **231**
Toxocara canis **218**
Toxoplasma gondii **190**
Toxoplasmosis **190**

Trachoma **69**
transactivator **145**
Traveler's diarrhea **37**
Treponema pallidum **61, 62**
Trichinella spiralis **217**
Trichinoses **217**
Trichomonas vaginalis **181**
Trichomoniasis **181**
Trichuris trichiura **210**
trigeminal ganglion **108**
Trophozoites **178**
Trypanosoma
    brucei **185**
    cruzi **184**
    gambiense **185**
    rhodesiense **185**
tsetse fly **185**
tubercle **81**
tubercle bacillus **81**
tuberculate macroconidia **169**
Tuberculin **82**
Tuberculoid leprosy **86**
Tularemia **55**
tumor necrosis factor (TNF) **80**
Tweety Bird **68**
type I alveolar cells **191**
Typhoid fever **39**
typhus fever **74**
    Endemic flea-borne **75**
    Epidemic louse-borne **74**
    murine **75**
    Scrub **75**
Tzanck **109, 111**
undulating fever **54**
Urease **41, 47, 172**
vaccine **249, 250**
Valley fever **167**

Index

Gram + Cocci | Gram – Cocci | Gram + Rods | Gram – Rods | Gram Nothing | DNA Viruses | RNA Viruses | Fungi | Parasites | X–Reference | Index

Index

The Microbiology Companion, Topf and Faubel ©1997

266

variable surface glycoprotein **185**
Variola
    major **104**
    minor **104**
vasovasorum **62**
VDRL **64**
venereal disease research laboratory **64**
Vibrio
    cholera **44**
    parahaemolyticus **44**
Vidarabine **109**
vif **145**
Viral capsid antigen **114**
Viral membrane antigen **114**
VZV immune globulin **111**
Waterhouse-Friderichsen syndrome (WFS)
  12
Weil-Felix reaction **41, 72**
West African sleeping sickness **185**
Western equine encephalitis **134**
whipworm **210**
Whooping cough **52**
window period **99**
wool sorters disease **17**
World Health Organization (WHO) **104**
Wound botulism **20**
Wright stain **187**
Wuchereria bancrofti **219**
Yellow fever **135**
Yersinia
    enterocolitica **56**
    pestis **56**
Ziduvudine (ZDV) **149**
Ziehl-Neelsen **82**
Zoster **111**
zygomycetes
    mucor **173**
    rhizopus **173**
Zygomycosis **173**

Section Six: Flash Cards

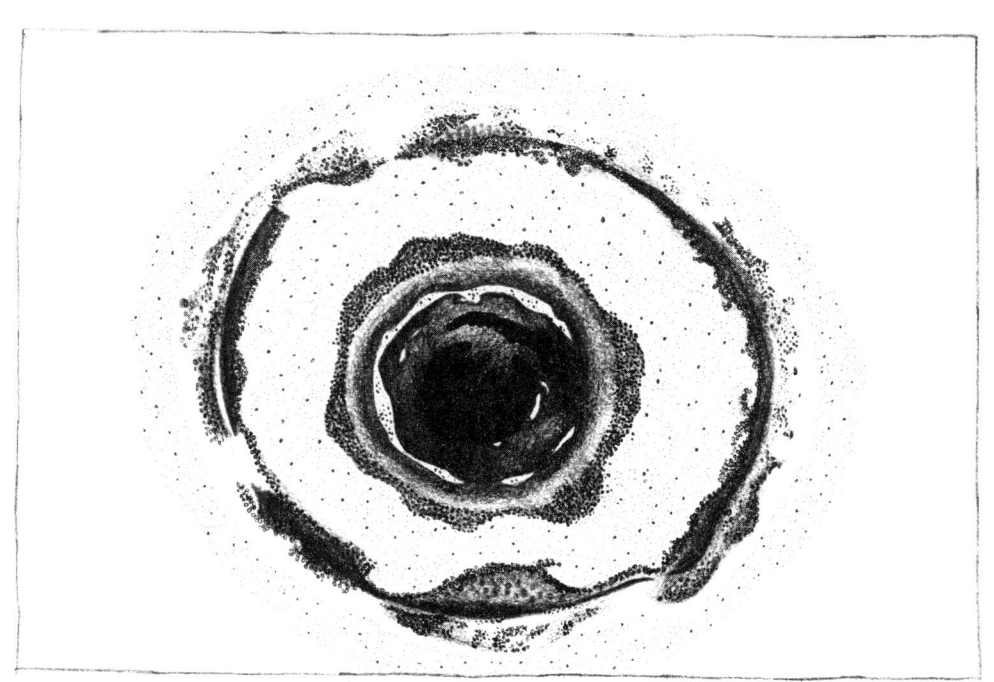

Herpes Virus, Dennis Franklin

Staphylococcus saprophyticus

Staphylococcus epidermidis

Staphylococcus aureus

Enterococcus faecalis faecium

group D

Streptococcus agalactiae

group B

Streptococcus pyogenes

group A

Streptococcus viridans

Streptococcus pneumoniae

pneumococcus

Streptococcus bovis

group D

G+ cocci, ß-hemolytic (clear), catalase +, coagulase +, facultative anaerobe.

Food poisoning (enterotoxin, heat stable), toxic shock syndrome (TSST-1), scalded skin syndrome (exfoliatin toxin), osteomyelitis (#1 cause).

Protein A (anti-complement and anti-phagocytic), alpha toxin (lysis of PMNs and platelets), ß-lactamase, altered penicillin binding proteins (confer MRSA).

G+ cocci, catalase +, coagulase -, facultative anaerobe, novobiocin sensitive.

Nosocomial infections, endocarditis.

Normal flora of skin.

Resistant to many antibiotics.

G+ cocci, catalase +, coagulase -, facultative anaerobe, novobiocin resistant.

Urinary tract infections.

Resistant to many antibiotics.

G+ cocci, ß-hemolytic (clear), catalase -, Lancefield group A, facultative anaerobe, bacitracin sensitive.

Pharyngitis (#1 bacterial cause), scarlet fever (erythrogenic toxin), pyoderma.

Non-suppurative sequelae: rheumatic fever, acute glomerulonephritis.

Pili, M protein (antiphagocytic), streptokinase, streptolysin O and S (cause hemolysis).

G+ cocci, ß-hemolytic (clear), catalase -, Lancefield group B, facultative anaerobe, bacitracin resistant.

Neonatal infections: early onset (within 48 hours, severe disease, pneumonia), late onset (within 10-60 days, milder disease, meningitis).

Postpartum endometritis.

Normal vaginal flora of 5-35% of women.

G+ cocci, α-hemolytic (green), catalase -, Lancefield group D, facultative anaerobe, bacitracin resistant, grows in 6.5% NaCl.

Secondary infection for UTI and wound infection, subacute bacterial endocarditis.

Normal flora of the intestines.

G+ cocci, a-hemolytic (green), catalase -, Lancefield group D, facultative anaerobe, bacitracin resistant, NO growth in 6.5% NaCl.

UTI, subacute bacterial endocarditis.

Normal flora of the intestines.

G+ cocci, α-hemolytic (green), catalase -, **no** Lancefield group, facultative anaerobe, inhibited by optochin, bile soluble.

Lobar pneumonia (#1 cause), meningitis, otitis media, sinusitis.

Capsule (anti-phagocytic, provides type specific immunity), IgA protease (facilitates colonization of upper respiratory tract).

Vaccine contains capsular antigens.

G+ cocci, α-hemolytic (green), catalase -, **no** Lancefield group, facultative anaerobe, **not** inhibited by optochin, **not** bile soluble.

Subacute bacterial endocarditis, dental carries.

Normal flora of the mouth.

Bacillus anthracis	**Neisseriaceae gonorrhoeae**	**Neisseriaceae meningitidis**
Clostridium tetani	**Clostridium botulinum**	**Bacillus cereus**
Corynebacterium diphtheriae	**Clostridium difficile**	**Clostridium perfringens**

G- diplococci, oxidase +, grows on Thayer-Martin medium and chocolate agar, ferments maltose.

Meningitis: maculopapular rash, epidemics in the military, spread by respiratory droplets.

Capsule (antiphagocytic), IgA protease.

Vaccine contains capsular antigens. Rifampin is used for prophylaxis.

G- diplococci, oxidase +, grows on Thayer-Martin medium and chocolate agar, does not ferment maltose.

Gonorrhea (STD), gonococcal arthritis, ophthalmia neonatorum.

Pili (antiphagocytic and allows colonization), membrane protein II, IgA protease.

Co-infection with chlamydia is common.

G+ rod, spores, catalase +, facultative anaerobe, needs O_2 to make spores.

Anthrax: cutaneous and pulmonary (wool sorters disease).

Anthrax toxin has 3 components: edema factor ($\uparrow$cAMP), lethal factor and protective antigen.

Capsule (anti-phagocytic) made of D-glutamate, all other proteins in nature are L-isomers.

G+ rod, spores, motile, aerobe

Food poisoning:
short incubation (1-6 hours, HS toxin, vomiting)
long incubation (10-12 hours, HL toxin ($\uparrow$cAMP), diarrhea).

Reheated fried rice a common source of infection.

G+ rod, spores, obligate anaerobe, motile.

Botulism: three types: food borne (spores germinate and produce toxin in anaerobic cans), infant (spores germinate and produce toxin in the baby's gut), wound (spores germinate and produce toxin in wound).

Botulism toxin blocks the release of acetylcholine at the neuromuscular junction. Symptoms: diplopia, weakness, flaccid paralysis, respiratory failure.

G+ rod, spores, obligate anaerobe, motile.

Tetanus (lockjaw) spores germinate and produce toxin in a wound.

Tetanus toxin: Coded by a plasmid. Prevents release of inhibitory neurotransmitters (GABA, glycine). Symptoms: spastic paralysis, respiratory failure.

Vaccine contains tetanus toxoid. Routine childhood vaccine: DTP.

G+ rod, spores, obligate anaerobe, nonmotile.

Food poisoning (enterotoxin) from eating spores. Diarrhea.

Gas gangrene (exotoxins): wound infection of necrotic, anaerobic tissue. Organism produces gas.

G+ rod, spores, obligate anaerobe, motile.

Pseudomembranous colitis: a necrotizing inflammatory lesion of the colon. Causes diarrhea. Associated with antibiotics (clindamycin and ampicillin especially).

2 enterotoxins: Toxin A causes diarrhea and Toxin B is cytopathic.

Treat with Flagyl and/or oral vancomycin.

G+ rod, no spores, club shaped, facultative anaerobe, grown on Loeffler medium.

Diphtheria: URI, may form a pseudomembrane which can obstruct the airway.

Diphtheria toxin stops protein synthesis by blocking elongation factor 2. (CDEF)

Vaccine contains diphtheria toxoid. Routine childhood vaccine: DPT.

Escherichia
coli

ETEC

Escherichia
coli

Listeria
monocytogenes

Escherichia
coli

EHEC

Escherichia
coli

EPEC

Escherichia
coli

EIEC

Klebsiella
pneumonia

Salmonella
typhi
cholerae suis
paratyphi

Shigella
dysenteriae,
flexneri,
boydii,
sonnei

G+ rod, ß-hemolytic (clear), motile, facultative anaerobe, intracellular reproduction.

Meningitis in infants, immunocompromised and renal transplant patients.

Neonatal infections: early onset (1st 48 hours, pneumonia, serious disease) late onset (1-4 wks, meningitis).

Can cause granulomas.

Transmission: contact with animals, animal products (milk), contaminated food, in utero, passage through birth canal.

G- rod, ferments lactose and glucose, facultative anaerobe, oxidase -, grows on MacConkey agar.

Diarrhea, UTI (#1 bacterial cause), meningitis in neonates, gram - septicemia.

Capsule (K), flagella (H), O-antigen (O) are used to categorize *E. coli*.

Pili allow colonization.

G- rod, ferments lactose and glucose, facultative anaerobe, oxidase -, grows on MacConkey agar.

Traveler's diarrhea: not invasive, 2 toxins: heat labile (LT) (A and B subunits, mechanism is similar to cholera toxin ($\uparrow$cAMP)

heat stable (ST) ($\uparrow$cGMP).

Colonizing factor antigen is a plasmid encoded pilus necessary for infection.

G- rod, ferments lactose and glucose, facultative anaerobe, oxidase -, grows on MacConkey agar.

Dysentery: fever, cramps and diarrhea with blood and pus; clinically identical to disease caused by *Shigella*.

Ability to invade the epithelial cells of the intestinal mucosa.

G- rod, ferments lactose and glucose, facultative anaerobe, oxidase -, grows on MacConkey agar.

Diarrhea and vomiting, stools contain mucous.

Unable to invade the intestinal epithelium.

EPEC adhesion factor: allows adhesion to intestinal mucosa, coded by a plasmid.

G- rod, ferments lactose and glucose, facultative anaerobe, oxidase -, grows on MacConkey agar.

Hemolytic colitis bloody diarrhea without pus. Transmission in hamburger meat.

O157:H7 most common serotype.

Shiga-like toxin I and II: inactivates ribosomes to stop protein synthesis; each toxin contains 6 subunits (one A and five Bs).

Destroys intestinal epithelium.

G- rod, does not ferment lactose, ferments glucose, nonmotile, H2S -, oxidase -, grows on MacConkey, methylene blue +.

Dysentery: bloody, mucopurulent diarrhea. Invades local mucosa does not cause bacteremia.

Transmission is by the 4 Fs.

Shiga toxin has 6 subunits (one A and five Bs): A subunit inactivates 60S ribosomes which stops protein synthesis. B subunit binds cells.

G- rod, does not ferment lactose, ferments glucose, motile, H2S+, oxidase -, grows on MacConkey, methylene blue +, produces H2S, intracellular.

Enterocolitis: high infective dose required, necrotizing inflammatory lesion of the colon. No bacteremia. **Typhoid fever**: systemic infection, can infect the gallbladder causing a chronic asymptomatic carrier state, rose spots.

Capsule: Vi. Flagella has phase variation.

G- rod, ferments lactose, ferments glucose, non-motile, quelling +, H2S+, urease +.

Pneumonia: nosocomial or opportunistic, thick currant jelly sputum. UTI, burn wound infections.

Capsule inhibits phagocytosis.

Coliform.

Proteus Providencia Morganella	**Serratia marcescens**	**Enterobacter cloacae**
Vibrio parahaemolyticus	**Vibrio cholera**	**Pseudomonas aeruginosa**
Helicobacter pylori	**Campylobacter jejuni**	

G- rod, lactose and glucose fermenter, motile, quellung -.

Most commonly associated with opportunistic and/or nosocomial infections.

UTI.

Coliform.

G- rod, slow lactose fermenter, ferments glucose, motile.

Most commonly associated with opportunistic and/or nosocomial infections.

Pneumonia: nosocomial and opportunistic.

Coliform.

G- rod, non-lactose fermenter, phenylalanine deaminase +, urease + (except *Providencia*), very motile, swarms over the agar.

UTI: community and nosocomial.

Urease increases pH of the urine: predisposes to renal stone formation, permits colonization of the urinary tract with bacteria.

Proteus induces antibodies which cross react with *Rickettsia*, basis of Weil-Felix reaction.

Coliform.

G- rod, obligate aerobe, does not ferment lactose, oxidizes glucose, oxidase +, motile.

Opportunistic infections: pneumonia, wound infections. **Nosocomial infections**: UTI, meningitis. **Swimmer's Ear** (otitis externa).

Exotoxin A blocks elongation factor 2 and stops protein synthesis.

Frequently contaminates water sources.

G- rod, facultative anaerobe, slow lactose fermenter, ferments glucose, oxidase +, no growth on 2% NaCl, grows at ↑pH.

Cholera: voluminous rice water stool, non-invasive, can be quickly fatal (within 8 hrs).

Enterotoxin (choleragen): 6 subunit toxin (one A and five Bs) which ↑cAMP. Mucinase permits adhesion to intestinal mucosa.

O-1 serotype causes epidemic cholera. non-O serotype causes sporadic cholera.

G- rod, facultative anaerobe, slow lactose fermenter, ferments glucose, oxidase +, growth on 8% NaCl.

Food poisoning: nausea, vomiting, diarrhea; transmission by eating raw or undercooked seafood. Common in Japan.

Toxin similar to choleragen.

G- rod, S-shaped, unable to grow on MacConkey, inert to glucose, oxidase +, nalidixic acid sensitive, motile.

Enterocolitis: abdominal pain and cramps, bloody diarrhea, disease is self-limited.

Enterotoxin is heat labile and ↑cAMP.

G- rod, S-shaped, unable to grow on MacConkey, inert to glucose, oxidase +, urease +.

Chronic gastritis and **duodenal ulcers**.

Urease raises the pH of the stomach so the pathogen can survive.

Fusobacterium	**Bacteroides melaninogenicus**	**Bacteroides fragilis**
Haemophilus ducreyi	**Haemophilus aegyptius**	**Haemophilus influenza**
Brucella *abortus* *melitensis* *suis*	**Bordetella pertussis**	**Legionella pneumophila**

G- rod, obligate anaerobe, grows on BAP with vancomycin and kanamycin.

Abscesses: pelvic and abdominal, secondary to trauma (surgery, IUD).

Capsule. No endotoxin activity.

Most common bacteria in the gut.

G- rod, obligate anaerobe, black pigment when grown on BAP.

Lung abscesses occur as a result of aspiration. Periodontal disease.

Capsule.

Normal flora of mouth, vagina and colon.

G- rod, obligate anaerobe, long slender rods tapered at both ends.

Sinus infections, ear infections, brain abscesses and lung infections. Often mixed with *B. melaninogenicus*.

Normal flora of mouth, vagina and colon.

G- coccobacillus, growth requires factor X (heme) and factor V (NAD) on chocolate agar.

Meningitis: #1 cause in children 6 months to 6 years. Epiglottitis, pneumonia, URI.

Capsule: antiphagocytic, 6 serotypes (type b causes most serious disease). IgA protease.

Vaccine contains type b capsular polysaccharide conjugated to a carrier protein. Routine childhood vaccine. Rifampin is used for prophylaxis.

G- coccobacillus, growth requires factor X (heme) and factor V (NAD) on chocolate agar.

Conjunctivitis (pink eye).

G- coccobacillus, growth requires factor X (heme) and factor V (NAD).

Chancroid: sexually transmitted disease characterized by soft painful ulcers.

G- rod, stain very lightly, cultures must be enriched with iron and cysteine, diagnoses is normally by titers.

Legionnaires' disease: atypical pneumonia, airborne transmission; no person-to-person transmission. Pontiac fever: milder infection.

ß-lactamase.

Intracellular reproduction in macrophages.

G- rod, grows on Bordet-Gengou medium, quellung +.

Whooping cough causes distinctive cough, spread by respiratory droplets.

Pertussis toxin: 2 subunits (A and B), causes irreversible activation of G-proteins. Capsule. Adenylate cyclase (inhibits PMNs). Pili.

Vaccine: contains killed organisms, is a routine childhood vaccine (DPT).

G- rod, aerobic, capsule.

Brucellosis: (undulating fever) animal infection transmitted to humans through occupational contact or ingestion of unpasteurized milk.

Intracellular reproduction with periodic release from cells in the RES causing the undulating symptoms.

Cause granulomas.

Yersinia enterocolitica	**Yersinia pestis**	**Francisella tularensis**
Treponema pallidum	**Leptospira interrogans**	genera of the family **Enterobacteriaceae**
Chlamydia pneumoniae	**Borrelia recurrentis**	**Borrelia burgdorferi**

G- rod, facultative anaerobe, diagnosis by serology. **Ticks** are the vector and **wild animals** are the reservoir. Transovarian transmission in ticks.

Tularemia is a tick borne disease, affects people who handle animals. Causes local ulcers, lymphadenopathy and **caseating granulomas**. Organism reproduces intracellularly in monocytes and PMNs.

Vaccine contains live attenuated organisms and is available to people at risk.

page 55

G- rod, oxidase -, grows on MacConkey, ferments glucose. **Fleas** are the vector and **rodents** are the reservoir.

The Plague: transmission is by fleas. Painful lymphadenopathy. Can cause pneumonia which results in respiratory transmission.

Capsule F1 (antiphagocytic), murine toxin, intracellular growth.

Vaccine available for people at risk.

family Enterobacteriaceae.

page 56

G- rod, oxidase -, growth on MacConkey, ferments glucose, motile.

Enterocolitis: fever and diarrhea; spontaneous recovery; arthritis is a late complication.

Heat stabile toxin ↑cGMP

family Enterobacteriaceae.

page 56

Escherichia coli
Shigella
Salmonella
Klebsiella
Enterobacter
Serratia
Proteus
Providencia
Morganella
Yersinia

page 34

Spirochete, thin, tightly coiled, motile, obligate aerobe, visible with darkfield microscopy, can be grown on artificial media.

Leptospirosis: transmitted by contact with infected urine, disease particularly affects the kidneys, liver, and meninges. Biphasic disease: acute and immune.

Dogs are the most frequent source of infection in the U.S.

page 60

Spirochete, tightly coiled, visible by darkfield microscopy.

Dx by Treponemal (FTA-ABS, MHA-TP) and indirect non-Treponemal tests (VDRL, RPR).

1° syphilis (STD) hard painless lesion.
2° syphilis rash 1-3 mo. after 1° lesion
3° syphilis 1-40 yrs after 1°: gummas, aneurysms and aortic regurgitation, tabes dorsalis, CNS, Argyll Robertson Pupil.

Congenital syphilis: often causes fetal death.

page 61

Spirochete, large, loosely coiled, motile, visible by light microscopy.

Ticks are the vector and small mammals and **deer** are the reservoir. Transovarian transmission in ticks.

Lyme disease: transmitted by bite of deer tick; initially causes fever, headache and a characteristic lesion: erythema chronicum migrans. Later causes arthritis and myocarditis.

page 58

Spirochete, large, loosely coiled, visible by light microscopy. Diagnosis by seeing pathogen in peripheral blood smear.

Human body **lice** are the vector and **humans** are the only reservoir.

Relapsing fever: transmitted by body lice. Most of the organism is cleared by antibodies to variable major protein (VMP). The organism then undergoes antigenic variation of VMP and the disease relapses (fever, chills, headache).

page 59

Obligate intracellular parasite, anaerobic, unable to make ATP, grown only in cell cultures. Inclusion bodies lack glycogen and stain with Giemsa.

Atypical pneumonia: common among college students.

Treat with erythromycin or tetracycline.

page 68

Rickettsiae rickettsii	**Chlamydia trachomatis**	**Chlamydia psittaci**
Rickettsiae typhi	**Rickettsiae prowazekii**	**Rickettsiae akari**
Mycoplasma pneumoniae	**Rickettsiae coxiella burnetii**	**Rickettsiae tsutsugamushi**

Obligate intracellular parasite, anaerobic, unable to make ATP, grown only in cell cultures. Sulfonamide sensitive.

Transmission by inhalation of organisms from bird feces.

Causes serious pneumonia.

Disease can spread to liver and thyroid causing jaundice and thyrotoxicosis. Can infect the meninges.

Obligate intracellular parasite, anaerobic, unable to make ATP, grown only in cell cultures. Sulfonamide sensitive.

Eye infections: **Trachoma** (type A, B, Ba, C), Inclusion conjunctivitis (type D-K).

Pneumonia: Infant pneumonitis (type D-K).

STD: Chlamydia(type D-K), Lymphogranuloma venereum (type L-1, L-2, L-3).

Obligate intracellular parasite, grown only in cell culture.

Ticks are the vector and **rodents** and **dogs** are the reservoir. Transovarian transmission in ticks.

Rocky Mountain Spotted Fever is transmitted by a tick, causes a rash which starts on the extremities and moves toward the trunk. Infects the endothelium of small blood vessels; causes vasculitis.

Obligate intracellular parasite, grown only in cell culture.

Mouse **mites** are the vector and **mice** are the reservoir.

Rickettsialpox transmitted by the bite of a mite. Causes a chickenpox-like lesion, recovery is spontaneous.

Obligate intracellular parasite, grown only in cell culture.

Human body **lice** are the vector and **humans** are the reservoir.

Epidemic louse-borne typhus fever: transmitted by body lice; organism grows in endothelium of vessels; rash starts on the trunk and moves to the extremities.

Brill-Zinsser disease is a mild sporadic form of typhus which occurs in people who have recovered from the epidemic form.

Obligate intracellular parasite, grown only in cell culture

Fleas are the vector and **rats** are the reservoir. Transovarian transmission in fleas.

Endemic flea-born typhus or murine typhus clinically identical to louse borne.

Obligate intracellular parasite, grown only in cell culture.

Rat **mites** are the vector and **rats** are the reservoir. Transovarian transmission in mites.

Scrub typhus: transmitted by mites, incubation is 1-2 weeks, causes fever headache and a lesion at the site of the mite bite.

Obligate intracellular parasite, grown only in cell culture. Organism can survive in the environment.

No arthropod vector.

Q fever: transmitted by inhalation of organism; fever, chills, atypical pneumonia. Does not cause a rash.

Smallest free living organism, no cell wall, ß-hemolytic (clear), cholesterol in cell wall and required for growth. Cold agglutinins in sera. Produces H_2O_2.

Atypical pneumonia in children and military recruits. Fever, rash and CNS symptoms. Humoral immunity is not protective and repeat infection is possible.

Treat with erythromycin or tetracycline (cell wall antibiotics are ineffective).

Mycobacteria tuberculosis	**Actinomyces asteroides**	**Actinomyces israelii**
Mycobacteria marinum	**Mycobacteria kansasii**	**Mycobacteria leprae**
Mycobacteria fortuitum complex	**Mycobacteria avium-intracellulare complex**	**Mycobacteria scrofulaceum**

G+, branching filamentous growth, obligate anaerobe.

Actinomycosis: sulfur granules (yellow masses of bacteria) form abscesses and draining sinuses.

Normal flora of oral cavity and GI tract.

Associated with surgery, trauma, IUDs.

page 79

Acid fast rod, branching filamentous growth, aerobe.

Nocardiosis: pneumonia in immunocompromised hosts. Causes abscesses and sinus tracts without sulfur granules.

Abscesses can form cavities like TB. If an abscess erodes a blood vessel the bacteria can spread hematogenously.

page 79

Acid fast rod, obligate aerobe, stain: Ziehl-Neelsen, grows on Löwenstein-Jensen media, cord factor, intracellular growth.

Tuberculosis. 1° TB induces cell mediated immunity, skin test +. Caseating granulomas. **2° TB** (reactivation) age, low immunity, stress. Lesions erode through bronchial walls, coughing spreads infection. **Miliary TB** infection spreads throughout the body.

BCG (live attenuated) vaccine is not used in the U.S.

pages 81-86

Acid fast rod, never been grown on artificial media, intracellular growth in skin histiocytes and Schwann cells.

Lepromatous leprosy is progressive and fatal (poor cellular immune response). Causes leonine facies, blindness, symmetric nerve damage. **Skin test -.**

Tuberculoid leprosy is self limiting due to normal cellular immune response, attacks the skin and causes anesthesia. **Skin test +.**

page 86

Acid fast rod, photochromogen (slow growth with yellow pigment when grown in light).

Cross reacts with PPD.

Pneumonia similar to TB.

Occurs most commonly in Texas.

page 88

Acid fast rod, photochromogen (slow growth with yellow pigment when grown in light).

Granulomatous ulcers form at abrasions and trauma.

Found in fresh, salt and swimming pool water.

Treat with tetracycline.

page 88

Acid fast rod, scotochromogen (slow growth with orange pigment when grown in light or darkness).

Scrofula: granulomatous inflammation of lymph nodes.

Treat by excision of nodes.

page 88

Acid fast rod, nonchromogen (slow growth with no pigment when grown in light or darkness).

Pulmonary infection identical to TB.

Is the most common systemic bacterial infection found in people with AIDS.

Found in the southeastern U.S.

High antibiotic resistance.

page 88

Acid fast rod, rapidly growing (rapid growth with no pigment when grown in light or darkness).

Wound infections following surgery, present as skin abscesses.

page 88

Enveloped DNA viruses	Non-enveloped DNA viruses	DNA viruses
HBV Ag and Ab in sera	HBV diseases	HBV characteristics
Smallpox virus	Poxviridae	Hepatitis D virus HDV

Non-enveloped
Papovavirus (DS, circular supercoiled, icosahedral)
Adenovirus (DS, linear, icosahedral)
Parvovirus (SS, linear, icosahedral)

Enveloped
Hepadnavirus (DS, circular incomplete, icosahedral)
Poxvirus (DS, linear, complex)
Herpes virus (DS, linear, icosahedral)

page 92

Non-enveloped
Papovavirus
    Papilloma
    JC
    BK
    SV40
    Polyomavirus
Adenovirus
    Adenovirus
Parvovirus
    B19

page 92

Enveloped
Hepadnavirus
    HBV (hepatitis B virus)
Poxvirus
    Smallpox virus
    Vaccina virus
    Molluscum contagiosum virus
Herpesvirus
    Herpes simplex I
    Herpes simplex II
    VZV (varicella-zoster virus)
    CMV (cytomegalovirus)
    EBV (Epstein-Barr virus)

page 92

Enveloped, icosahedral, DS circular incomplete DNA. DNA dependent DNA polymerase, RNA-dependent DNA polymerase.

Life cycle: includes reverse transcriptase, envelope is acquired by budding though the cell membrane. Sera of infected persons contains HBsAg, Dane particles, and HBeAg.

Transmission: primarily through blood, (needle sticks, IV drug use), mother to child (passage through birth canal, breast feeding), sexual.

page 95

Acute hepatitis: clinically similar to hepatitis A, long incubation period (2-6 months). Symptoms: fever, rash, arthritis and jaundice. May be self-limited or progress to chronic hepatitis.

Chronic hepatitis: persistence of HBsAg in sera longer than 6 months, occurs in 10% of HBV infections. 1. Chronic persistent: minimal necrosis. 2. Chronic active: more severe, piecemeal necrosis, bridging necrosis, cirrhosis, primary hepatocellular carcinoma.

page 97

HBsAg	**HBsAb**
1st antigen in sera contained in vaccine	appears in about 6 mo. indicates recovery
HBcAg	**HBcAb**
no serologic test	1st antibody in sera + during window phase
HBeAg	**HBeAb**
indicates transmissibility	2nd antibody in sera indicates low transmissibility

page 99

Small SS, circular RNA. Defective virus. Envelope is HBsAg.

Co-infection: simultaneous infection of HBV and HDV. Indistinguishable from HAV or HBV alone. Usually resolves.

Super-infection: infection with HDV occurs in someone with chronic HBV. Hepatitis more severe, chronic HDV infection, increased risk of cirrhosis.

page 101

DS linear DNA, brick shaped, complex nucleocapsid, enveloped (double membrane). DNA dependent RNA polymerase.

Life cycle: all steps occur in the cytoplasm. Envelope not acquired by budding through cell membrane.

Smallpox virus
Vaccina virus
Molluscum contagiosum virus

page 103

Smallpox: only disease that has been eradicated. One serotype, no animal reservoir, no asymptomatic or latent infections.

Symptoms: fever, malaise. Rash: papules-> vesicles-> pustules-> crusts.

Vaccine contains live vaccina virus. No longer routinely administered.

page 104

Herpes simplex virus I

HSV-I

Herpes

viruses

Vaccina virus

Molluscum contagiosum
virus

Cytomegalovirus

CMV

Varicella-zoster virus

VZV

Herpes simplex
virus II

HSV-II

Polyoma
viruses

Papovaviridae

Epstein-Barr virus

EBV

Poxviruses.

Vaccina virus: non-pathogenic, origin unknown, live virus used for smallpox immunization.

Molluscum contagiosum virus: causes a cutaneous wart-like infection; transmission by direct contact with lesions. Lesions are self-limited.

Herpes simplex I

Herpes simplex II

Varicella-zoster virus

Cytomegalovirus

Epstein Barr virus

Human Herpes Virus 6–8

Herpesvirus. DS linear DNA, icosahedral, envelope.

Latent in sensory ganglia of head.

Gingivostomatitis (1°), **herpes labialis** (2°), herpetic whitlow, keratoconjunctivitis, encephalitis. Cell mediated immunity important for recovery.

Tzanck smear for diagnosis. Acyclovir for treatment.

Herpesvirus. DS linear DNA, icosahedral, envelope.

Latent infections in sensory ganglia of lumbar and sacral regions.

Genital herpes (1° and 2°), neonatal herpes, cervical carcinoma.

Tzanck smear for diagnosis. Acyclovir for treatment.

Herpesvirus. DS linear DNA, icosahedral, envelope.

Latent in sensory ganglia, dorsal root ganglia of spinal cord.

Varicella (chickenpox) 1°. **Reye's syndrome** is associated with aspirin.

Zoster (shingles) 2°.

Tzanck smear for diagnosis. Acyclovir for treatment.

Herpesvirus. DS linear DNA, icosahedral, envelope. One serotype.

Latent in leukocytes.

Cytomegalic inclusion disease in fetus, transmitted across the placenta. Most common congenital virus infection.

Heterophile negative mononucleosis.

Persons receiving organ transplants are prone to CMV.

Herpesvirus. DS linear DNA, icosahedral envelope.

Latent infection in B-cells.

Heterophile-positive infectious mononucleosis: transmitted by saliva, large atypical lymphocytes found in sera. Sore throat, hepatosplenomegaly.

Burkitt's lymphoma: lymph node malignancy of lower jaw, common in Africa.
Nasopharyngeal carcinoma, **chronic fatigue syndrome.**

DS circular DNA, icosahedral, no envelope.

Human **pa**pilloma virus

Mouse **po**lyoma virus

Simian **va**cuolating virus (SV40)

Human polyoma virus (JC virus)

Human polyoma virus (BK virus)

DS circular DNA, icosahedral, no envelope.

Mouse polyoma virus: causes malignant tumors to develop when injected into mice.

SV40: malignant transformation in rodent and human cells.

JC virus: progressive multifocal leukoencephalopathy.

BK virus: discovered in a renal transplant patient.

Parvovirus **B19**	**Adenovirus**	**Human papilloma virus** **(HPV)**
Paramyxoviruses	**Orthomyxovirus** **(Influenza A, B and C)**	**RNA enveloped viruses**
Respiratory syncytial virus **(RSV)**	**Mumps virus**	**Measles virus**

DS circular DNA, icosahedral, no envelope.

Common warts: HPV 1-4, 10, 26, 28.

Condylomata acuminata (genital warts): STD, caused by HPV 6, 11 and 30.

Cervical carcinoma: HPV 16, 18 and 31.

Epidermodysplasia verruciformis: inherited disease allows severe wart infection; warts commonly become malignant.

DS linear DNA, icosahedral, no envelope. Hemagglutinin fiber which binds RBCs.

Respiratory infections, epidemic keratoconjunctivitis, gastroenteritis, hemorrhagic cystitis.

Not associated with cancer.

Vaccine: live (non-attenuated) virus for military.

SS linear DNA, icosahedral, no envelope. Smallest icosahedral virus.

Erythema infectiosum (Fifth disease): slapped cheeks.

Attacks erythroid precursor cells: aplastic crisis, chronic anemia.

Spontaneous abortion.

SS (-) RNA, not segmented, helical, envelope.
Orthomyxovirus: influenza A, B and C

Paramyxovirus: measles, mumps, RSV, parainfluenza

Togavirus: rubella, alpha (EEEV, WEEV)

Flavivirus: St. Louis encephalitis virus, yellow fever, dengue fever virus, hepatitis C

Rhabdovirus: rabies virus

SS segmented (-) RNA, helical, envelope with H (hemagglutinin) and N (neuraminidase).

H and N are type specific antigens.
Antigenic shift: major change, infrequent.
Antigenic drift: minor change, frequent.

Influenza. Reye's syndrome is associated with taking aspirin.

Amantadine for influenza A.

Vaccine contains influenza A and B. Yearly vaccines to keep up with drift and shift.

Measles: H and F

Mumps: HN and F

RSV: F

Parainfluenza: HN and F

H: hemagglutinin
N: neuraminidase
F: fusion protein (the fusion proteins of measles and mumps are also hemolysins).

Paramyxovirus. SS (-) RNA, not segmented, helical, envelope with H and F (also a hemolysin). RNA dependent RNA polymerase.

Measles: Koplik's spots, rash with multinucleated giant cells. **Encephalitis, subacute sclerosing panencephalitis.**

Vaccine contains live attenuated virus. Routine childhood vaccine (MMR).

Paramyxovirus. SS (-) RNA, not segmented, helical, envelope with HN and F (also a hemolysin). RNA dependent RNA polymerase.

Mumps: transmitted by respiratory droplets with swelling of the parotids, orchiditis, meningitis.

Vaccine contains live attenuated virus. Routine childhood vaccine (MMR).

Paramyxovirus. SS (-) RNA, not segmented, helical, envelope with F. Fusion protein fuses cells forming multinucleated syncytia. RNA dependent RNA polymerase.

Lower respiratory tract infections (pneumonia and bronchiolitis) occur in infants and may be severe. **Upper respiratory tract infections** occur in children and adults.

Recovery from infections does not induce immunity. No vaccine.

Rubella virus	Togaviruses	Parainfluenza virus
Flaviviruses	Western equine encephalitis virus	Eastern equine encephalitis virus
Dengue fever virus	Yellow fever virus	St. Louis encephalitis virus

Paramyxovirus. SS (-) RNA, not segmented, helical, envelope with HN and F. RNA dependent RNA polymerase.

Lower respiratory tract infections occur in children. **Croup** (laryngotracheobronchitis): inspiratory and expiratory stridor, bark-like cough.

Upper respiratory infections occur in adults: pharyngitis, colds.

No vaccine.

SS (+) RNA (infectious genome), icosahedral, envelope with hemagglutinin protein spikes.

Rubella virus

Alpha virus

  Eastern equine encephalitis virus

  Western equine encephalitis virus

Togavirus. SS (+) RNA (infectious genome), icosahedral, envelope with hemagglutinin protein spikes.

Rubella: mild illness with maculopapular rash.

Congenital rubella syndrome: serious illness which occurs in the fetus when the mother is infected. Affects head, heart, ears.

Vaccine contains live attenuated virus. Routine childhood vaccine (MMR).

Togavirus. Arbovirus. SS (+) RNA (infectious genome), icosahedral, envelope with hemagglutinin protein spikes.

Eastern equine encephalitis is characterized by sudden onset of headache, nausea, vomiting and fever. High mortality rate.

Mosquitoes are the vector and **wild birds** are the reservoir. Humans and horses are dead end hosts.

Togavirus. Arbovirus. SS (+) RNA (infectious genome), icosahedral, envelope with hemagglutinin protein spikes.

Western equine encephalitis is similar to but less severe than eastern equine encephalitis.

Mosquitoes are the vector and **wild birds** are the reservoir. Humans and horses are dead end hosts.

Arbovirus. SS (+) RNA (infectious genome), icosahedral, envelope. Smaller than the other togaviruses.

St. Louis encephalitis virus

Yellow fever virus

Dengue fever virus

Hepatitis C virus

Flavivirus. Arbovirus. SS (+) RNA (infectious genome), icosahedral, envelope.

St. Louis encephalitis is an arbovirus infection which occurs in urban areas.

Mosquitoes are the vector and **wild birds** are the reservoir.

Flavivirus. Arbovirus. SS (+) RNA (infectious genome), icosahedral, envelope.

Yellow fever is an acute infection which causes liver damage and jaundice.

Vaccine contains live attenuated virus and is given to travelers and persons living in endemic areas.

Mosquitoes are the vector and **monkeys** and **humans** are the reservoir.

Flavivirus. Arbovirus. SS (+) RNA (infectious genome), icosahedral, envelope. Four cross reacting antigenic types.

Dengue fever (syn. break-bone fever) is characterized by headache fever and severe pain in the joints and muscles. Two complications:

  Dengue hemorrhagic fever
  Dengue shock syndrome

Mosquitoes are the vector and **monkeys** and **humans** are the reservoir.

RNA non-enveloped viruses	Rabies virus (a rhabdovirus)	Hepatitis C virus
Sabin vaccine	Salk vaccine	Poliovirus
Hepatitis A virus	Echovirus	Coxsackievirus

Flavivirus-like. SS RNA, envelope.

Hepatitis occurs through contact with infected blood. Is the most common cause of post-transfusion hepatitis.

Acute hepatitis is identical to the hepatitis caused by HBV and HAV.

Chronic hepatitis is more likely than with HBV but it is less severe.

SS (-) RNA, bullet shaped, envelope, RNA dependent RNA polymerase. Infected neurons contain **Negri bodies**.

Rabies is transmitted by the bite of an infected animal. Confusion, lethargy, hydrophobia.

Vaccine (HDCV) for persons at risk contains inactivated (killed) virus. HDCV and rabies immune globulin are given at different sites for postexposure prevention (HDCV is given four more times).

Picornaviruses

Enteroviruses
    Poliovirus
    Coxsackievirus
    Echoviruses
    Hepatitis A

Rhinoviruses

Reoviruses

Rotavirus

Orbivirus

Picornavirus, enterovirus. SS (+) RNA (infectious genome) Protein on 5' end (acts as promoter), icosahedral, no envelope.

Poliomyelitis: fecal-oral transmission, replicates within anterior horn motor neurons resulting in paralysis.

Abortive polio myelitis

Nonparalytic polio myelitis

Paralytic polio myelitis

Salk Vaccine

• **k**illed virus

• all three serotypes

• given by injection

• induces IgG in the blood

• does not induce secretory IgA

• no possibility of reversion

• does not need refrigeration

Sabin Vaccine (TOPV)

• live attenuated virus

• all three serotypes

• given orally

• induces IgG and secretory IgA

• replicates in the GI and is shed in the feces, providing community inoculation

• reversion is possible and does occur, presently, in the U.S, reversion of vaccine is the most common cause of polio

• must be kept refrigerated

Picornavirus, enterovirus. SS (+) RNA (infectious genome). Protein on 5' end (acts as promoter), icosahedral, no envelope. Group A and B determined by pathogenicity in mice.

Life cycle similar to poliovirus.

Aseptic meningitis caused by group A and B.

Group A infections: herpangina, hand-foot and mouth disease.

Group B infections: pleurodynia, myocarditis.

Picornavirus, enterovirus. SS (+) RNA (infectious genome). Protein on 5' end (acts as promoter), icosahedral, no envelope.

Life cycle similar to poliovirus.

Aseptic meningitis: fecal-oral transmission.

Infantile diarrhea.

Picornavirus, enterovirus. SS (+) RNA (infectious genome). Protein on 5' end (acts as promoter), icosahedral, no envelope.

Acute hepatitis: fecal oral transmission, short incubation (20-25 days), no chronic infection or carrier state.

Retroviruses   Rotavirus   Rhinovirus

Systemic   Human   Human T-cell leukemia
fungi   immunodeficiency virus   virus

HIV   HTLV

Coccidioides   Cutaneous   Opportunistic
immitis   fungi   fungi

Picornavirus, enterovirus. SS (+) RNA (infectious genome). Protein on 5' end (acts as promoter), icosahedral, no envelope.

The common cold: upper respiratory tract infection (cough, sneeze, runny nose); transmission by contact with respiratory droplets off surfaces; rhinovirus is most common cause. Treatment is symptomatic.

page 158

Reovirus. DS segmented (11 pieces) RNA, icosahedral (double layered), no envelope, 4 serotypes, RNA dependent RNA polymerase.

Gastroenteritis: fecal-oral transmission; nausea, vomiting, watery diarrhea; common in infants.

page 160

Diploid SS RNA, icosahedral, envelope. Integrase, reverse transcriptase (RNA dependent DNA polymerase).

Human T-cell leukemia virus (HTLV)

Human immunodeficiency virus (HIV)

page 143

Retrovirus (diploid SS RNA), icosahedral, envelope. Integrase, reverse transcriptase (RNA dependent DNA polymerase). Tax and rex genes.

Leukemia and **lymphomas**: T-cell malignancies.

Tropical spastic paralysis: possible association with multiple sclerosis.

page 153

Retrovirus (diploid SS RNA), icosahedral, envelope. Integrase, reverse transcriptase (RNA dependent DNA polymerase). Tat, rev, and nef genes.

AIDS: opportunistic infections and malignancies, Kaposi's sarcoma, AIDS dementia complex.

Transmission via bodily fluids. Virus infects cells containing the surface molecule CD4 (T-cells, macrophages, monocytes). The envelope protein gp 120 of the virus binds to CD4.

page 146

All systemic fungi are dimorphic.

Coccidioides immitis

Histoplasma capsulatum

Blastomyces dermatitidis

Paracoccidioides brasiliensis

page 151

No opportunistic fungi is dimorphic.

Candida albicans   (yeast only)

Cryptococcus neoformans   (yeast only)

Aspergillus fumigatus   (mold only)

Zygomycetes   (mold only)

  Mucor

  Rhizopus

page 166

All cutaneous fungi are dimorphic.

Dermatophytes

    Epidermophyton

    Microsporum

    Trichophyton

Malassezia furfur

Cladosporium werneckii

Sporothrix schenckii

page 166

Systemic fungus. Dimorphic. Should not be cultured due to danger to lab workers.

Southwestern U.S.

Coccidioidomycosis: transmission by inhalation of arthrospores. Dissemination causes osteomyelitis and meningitis.

Valley fever: hypersensitivity reaction to *Coccidioides* infection; arthralgia, cough, fever, and erythema nodosum.

page 167

Paracoccidioides brasiliensis

Blastomyces dermatitidis

Histoplasma capsulatum

Aspergillus fumigatus

Cryptococcus neoformans

Candida albicans

Malassezia furfur

Dermatophytes
Epidermophyton
Microsporum
Trichophyton

Zygomycetes

Mucor
Rhizopus

Systemic fungus. Dimorphic: mold in soil, septate hyphae; yeast in the body. Sexual spores: microconidia, tuberculate macroconidia.

Mississippi and Ohio River valleys.

Histoplasmosis: transmission by inhalation of microconidia in bird droppings. Caseous granulomas form in the lung, delayed hypersensitivity skin test. Commonly asymptomatic, disseminated infection in young, old and immunocompromised. Reactivation of latent infection may occur.

Systemic fungus. Dimorphic: mold in the soil; yeast in the body, yeast is pear shaped with a broad based bud.

Southeastern and Midwestern U.S.

Blastomycosis (Gilchrist's disease): transmission by inhalation of conidia, pulmonary disease may progress to pneumonia.

Disseminated infection causes ulcerated granulomas in the skin and bones.

Systemic fungus. Dimorphic: mold is found in the soil and appears filamentous; yeast is found in the body.

Central and South America.

Paracoccidioidomycosis: transmitted by inhalation of conidia, most infections are asymptomatic.

Disseminated infection may cause ulcerations of lymph nodes and oral and nasal mucosa.

Opportunistic fungus. Not dimorphic. Yeast only. Reproduces by budding. Pseudohyphae in tissue. Asexual spores: chlamydospores.

Candida is part of the normal flora of the vagina, gut and oral cavity.

Candidiasis. Oral (thrush) often occurs in children. Vaginitis occurs most commonly in women who are diabetic, pregnant, or taking antibiotics or oral contraceptives.

Diaper rash, esophagitis.

Opportunistic fungus. Not dimorphic. Yeast only. Capsule. Diagnosis by **India ink** test.

Cryptococcosis: transmission by inhalation of conidia in pigeon and other bird droppings. Opportunistic infection (common in AIDS patients). Infection is characterized by a lack of a tissue response (no inflammation).

Meningitis.

Opportunistic fungus. Not dimorphic. Mold only. Septate hyphae appear in tissue.

Lung infections: allergic aspergillosis (eosinophilia and exudate), farmer's lung (allergic bronchospasm), fungus balls (grows in lung cavities), pneumonia.

Disseminated infection may cause granuloma formation throughout the body.

Opportunistic fungus. Not dimorphic. Molds are saprophytic and found in soil. Nonseptate, broad hyphae found in tissue.

Zygomycosis: immunocompromised (diabetic ketoacidotics) at risk.
Pulmonary infection: pneumonia. Rhinocerebral disease: transmission by inhalation of organism which invades nasal mucosa and proliferates in blood vessels; organism continues to grow upward eventually reaching the brain.

Cutaneous fungus. Dimorphic. Produce asexual conidia.

Infections occur in superficial keratinized structures. **Ringworm**. Athlete's foot, jock itch.

Treatment with topical agents or griseofulvin.

Dermatophytid reactions.

Cutaneous fungus. Dimorphic: grows primarily as a yeast.

Tinea versicolor: characterized by light lesions. Diagnosis by observation of budding yeast cells and hyphae together in prepared skin scrapings.

**Strongyloides
stercoralis**

**Necator
americanus**

New World Hookworm

**Ancylostoma
duodenale**

Old World Hookworm

**Toxocara
canis**

**Ancylostoma
braziliense**
Cat Hookworm

**Ancylostoma
caninum**
Dog Hookworm

**Trichinella
spiralis**

Intestinal nematode. 4 teeth to attach to the intestines, 1 cm long, "S" shaped.

Adults live in the *intestines* and feed on blood. The **eggs** are passed in *stool* and become free living **rhabditiform larvae**. The larvae then change to **filariform larvae** and penetrate human skin. The larvae travel into the *lung* to enter the alveoli. They are coughed up and swallowed.

Ground itch, pneumonitis, microcytic anemia.

page 212

Intestinal nematode. Knife-like fins (cutting plates) attach to intestines, 1 cm long, "S" shaped.

Adults live in the *intestines* and feed on blood. The **eggs** are passed in *stool* and become free living **rhabditiform larvae**. The larvae then change to **filariform larvae** and penetrate human skin. The larvae travel into the *lung* to enter the alveoli. They are coughed up and swallowed.

Ground itch, pneumonitis, microcytic anemia.

page 212

Intestinal nematode. Smallest intestinal nematode: 3 mm long.

3 kinds of life cycles:
Hookworm: life cycle identical to hookworms except the eggs hatch while in the body and noninfectious larvae are released in the stool.
Autoinfection: the larvae mature into the infectious form while still in the body.
Free-living: larvae mature into adult worms outside of the body.

Strongyloidiasis: skin, lung and intestinal symptoms.

page 213

Tissue nematode. **Adults** live in the *small intestines*. Females produce **eggs** which hatch into **larvae** while in the uterus. The larvae enter the circulation and encyst in *striated muscle*. When the host is eaten by another predator the larvae mature.

Trichinosis: transmission by eating undercooked meat (pork). Fever, muscle pain, muscle weakness, CHF, respiratory failure.

Humans are dead-end hosts.

page 217

Tissue nematode. **Adults** live in the *intestines* of *cats* or *dogs*. The **larvae** are passed in stool and infect *humans* by penetrating the skin. The worm cannot complete its life cycle and just travels through the *subcutaneous tissue*.

Cutaneous larva migrans: red itchy rash in a linear pattern. **Loeffler's syndrome**: eosinophilia and transient migratory pulmonary infiltrates.

Humans are dead-end hosts.

page 218

Tissue nematode. **Adults** live in the intestines of *dogs*, *wolves*, *foxes*. **Eggs** are passed in stool. When eggs are eaten by dogs they mature into **larvae**, enter blood stream, pass into *alveoli*, are coughed-up and swallowed. In humans, larvae cannot enter alveoli, they enter other tissues and cause damage.

Visceral larva migrans most often damages the liver, lungs heart and muscles. It causes necrosis, bleeding and granulomas.

Humans are dead-end hosts.

page 218

**Dracunculus
medinensis**

Guinea fire worm

**Brugia
malayi**

**Wuchereria
bancrofti**

Intestinal nematodes

**Loa
loa**

African eye worm

**Onchocerca
volvulus**

River blindness

Tissue nematodes

Tissue nematode. Microfilariae present in systemic circulation at night. **Humans**: only definitive hosts; **mosquitoes**: intermediate hosts.

Bancroftian filariasis: transmission by mosquitoes (larvae injected into skin). Symptoms from presence of worms in the lymphatics.

Elephantiasis: occurs after repeat infections, swelling of legs and genitals from obstruction of lymphatics.

page 219

Tissue nematode. Life cycle identical to *Wuchereria bancrofti*. Morphologically distinct from *Wuchereria bancrofti*.

Humans and other mammals: definitive hosts. **Mosquitoes** are the intermediate hosts.

Malayan filariasis: clinically identical to bancroftian filariasis.

page 221

Tissue nematode. Adults live in subcutaneous tissue. Females grow up to 1 meter. Humans: definitive hosts; copepods: intermediate hosts.

A portion of the female's body ulcerates through the skin surface and releases **larvae** when underwater. The larvae infect *crustaceans* and mature. *Humans* drink water with crustaceans. Larvae enter the *lymphatics* and *subcutaneous tissues* and mature into adults.

Dracunculiasis. Skin manifestations: ulcers and blisters. Systemic symptoms from hypersensitivity: vomiting, diarrhea.

page 220

Tissue nematode. Adults live in the subcutaneous tissues in areas called dermal nodules. **Humans**: only definitive hosts; **black flies**: intermediate hosts.

Onchocerciasis: transmission is by the bite of the black fly. The adult produces microfilariae which can migrate in the subcutaneous tissues for 2 years. Can result in blindness when the microfilariae concentrate in the eyes. Dermal nodules are non-tender, hard, mobile and 1-3 cm in diameter.

page 221

Tissue nematode. Adults live in the subcutaneous tissues all over the body. **Humans** and monkeys: definitive hosts; **deer flies**: intermediate hosts.

Loiasis: transmission is via a dear flies. Clinical symptoms result from the migration of the adult which causes localized allergic reactions called Calabar swellings. The adult worm can migrate across the eye.

page 222

Enterobius vermicularis

Trichuris trichiura

Ascaris lumbricoides

Hookworms

  Ancylostoma duodenale

  Necator americanus

Strongyloides stercoralis

page 208

Trichinella spiralis

Ancylostoma braziliense

Ancylostoma caninum

Toxocara canis

Wuchereria bancrofti

Brugia malayi

Dracunculus medinensis

Onchocerca volvulus

Loa loa.

page 216